HOW TO DIE IN SPACE

HOW TO
DIE
IN SPACE

A JOURNEY THROUGH DANGEROUS
ASTROPHYSICAL PHENOMENA

PAUL M. SUTTER, PhD

PEGASUS BOOKS
NEW YORK LONDON

HOW TO DIE IN SPACE

Pegasus Books Ltd.
148 W. 37th Street, 13th Floor
New York, NY 10018

First Pegasus Books cloth edition June 2020

Interior design by Maria Fernandez

Library of Congress Cataloging-in-Publication Data is available.

ISBN: 978-1-64313-438-3

10 9 8 7 6 5 4 3 2 1

Printed in the United States of America
Distributed by Simon & Schuster
www.pegasusbooks.us

To the mothers, mine and yours:
We must venture into the endless void
Never at rest, facing the dangers of our universe
Thank you, for teaching us how home should feel.

—Dedication, Rhyme of the Ancient Astronomer

CONTENTS

PROLOGUE
A Disclaimer

But which is worse
to die lost and alone
or surrounded by the people
you were trying to leave behind?
—Rhyme of the Ancient Astronomer

You're not going to make it in space.

I said, no.

Look, just because you're a child of Mother Nature, it doesn't mean she has to love you.

She can pull you below an event horizon, never to be seen by a living soul again. She can slam a mountain into you at ten thousand miles per hour, smashing you into dust. She can bore you to death, forcing you to spend eon after endless eon just to hop to the next star system. She can even *microwave* you. Literally cook you with microwaves. She can dose you with so much radiation that if you're supremely lucky you'll *only* get aggressive cancer. She can . . . you get the idea.

Space is nasty.

It's a rough universe out there, and I'm surprised that anybody, let alone you, would want to explore it. Sure, it's full of wonders: lacy tendrils of gas stretching for light-years, stellar explosions that can be seen from across the universe, dead and decaying stars filled with matter in the strangest of states, the list goes on. A beautiful and wonderful cosmos, full of colors, motion, and vitality. Staggeringly big, room enough for everyone. Full of enough surprises and mysteries to satisfy generations' worth of curiosity-seekers.

Lured by the wonders, eager young explorers go out ill-equipped and unprepared. They go off to hunt for the strange, the unique, the exotic. To dance through nebulae and surf on waves of gravity. To attempt to fathom the most closely held secrets of nature. To go and go and go, never looking back. Hundreds of billions of stars in every galaxy; hundreds of billions of galaxies in the observable universe.

They go to see the stars. Factories of fusion. Fountains of creation. Watchful guardians of the deep.

They go to see the nebulae. Tombs of the fallen. Birthplaces of light. Forges of the elements.

They go to see the unseeable. Whispers of distant collisions. Secrets written in strange matter. The great expanse of nothing.

They go to see the extreme. Gateways to new universes. Artifacts of the ancient cosmos. A new friend.

They go to see. To explore, to study, to observe, to witness.

They meet their ends much too soon. Caught in the gravitational pull of a black hole. Struck by a rogue comet. Blasted with radiation from the outburst on the surface of a star. Tragedies, all of them. Senseless and unnecessary.

So here I am. My first priority is to warn you off the whole escapade altogether. Find a planet, find a rock, call it home. Raise a farm. Raise some kids. You can't get rid of all the dangers in your life, but you sure can avoid the most obvious ones. Put some dirt under your feet and some air over your head. Get yourself a nice steady star with billions of years left of heat and light and warmth, and a nice steady planet with plenty of liquid water. Get a hobby, and get your mind off space.

Buy a telescope. Enjoy it from afar.

But you're not going to listen to me, are you? You're going to go there, aren't you? You're not like the others. You're not the stupid one, or the ignorant one, or the lazy one. You'll be clever and watchful and careful. You'll come back home with tales of wonder and awe.

You think you have one up on Mother Nature, do you? Just remember that she has a few billion years of experience.

So, second priority. If you're not going to stay put, I might as well tell you about some of the dangers you'll be facing. I'll assume you've solved the simple stuff, like how to actually get up into space, how to bring enough food and water and air with you, and how to navigate and travel. That's all just engineering problems, really, and not my department.

My department is physics—astrophysics. That means figuring out how stuff works up in space. And I'll be spilling a lot of astrophysics all over your nice clean new dreams. Sometimes I'll be brief, and sometimes I'll need to slow down and dig into the dirt. This isn't just a list of hazards but an explanation of *why* they're hazardous.

I would prefer that you end up not only alive, but also smart.

What I'm writing represents the latest scientific knowledge acquired from decades, and in some cases centuries, of research from Earth scientists. That means that a good chunk of it is right, but some of it might be wrong. That's just the way it is. I'll try my best to let you know when something's known for sure and when something's a little bit suspect—or even downright speculative. Again, use your judgment. I recommend treating everything I say as the Gospel Truth, at least for safety's sake.

You can never be too careful, out there in the void.

This will not be an exhaustive list, either. I have a deadline to write this, after all, and I can't wait around for every new discovery or piece of knowledge to make its way into these chapters. I'll hit the most obvious dangers and a few lesser known second-stringers. There are of course more threats out there, and as smart as I am I'm not omniscient. That's just the way the universe works, pal.

Most importantly, and I can't say this strongly enough: I will not be held liable for any inaccuracies, mistakes, or incomplete knowledge in the following pages. I will of course strive to minimize all such things, but nobody's perfect. Even me.

Travel at your own risk. If I say a star in a particular stage of evolution should be stable for another million years, and instead it goes supernova, don't blame me, blame physics. The universe is a complicated place, and the physics I'm about to describe isn't always simple.

I don't know how far you'll get or what you'll eventually encounter in your adventures. Our universe is in a constant state of flux; a big, messy existence. Things will catch you off guard. The cosmos will surprise you.

You have been warned.

So let us begin.

PART ONE

INTERPLANETARY THREATS

The Vacuum

Pulling for breath
but nothing comes.
Your eyes are weary, skin cold
heart beating in betrayal.
—Rhyme of the Ancient Astronomer

The main problem with space—and the first problem that you'll have to contend with as soon as you leave your precious atmospheric bubble—is that it's filled with nothing. Now, normally "nothing" would be a godsend, especially since as we continue in our journey we're going to encounter all manner of not-nothings that can slice you, irradiate you, or just plain smash you, so you might be surprised to learn that *no really, I mean nothing at all* is one of the most dangerous parts of this whole enterprise.

But that's part of the wonder of exploring the universe. Just when you think the cosmos has run out of things to harm you with, it turns to its ultimate ace up the sleeve, which isn't even a card at all. It's nothing.

It's vacuum. Nothing. No thing.

The French call it *sous vide*, which an Internet translator tells me directly translates to "under empty," as if you could take the emptiest thing possible—say, by pouring the last drop of milk out of the jug—rendering it completely devoid of liquid, and make it somehow even *more* empty. I wonder if the name came from the early days of making a vacuum—remove the air from a chamber as hard as you can, and keep going because there's got to be at least one more molecule in there.

"Is it empty yet?"

"Oui!"

"Not good enough, monsieur. Make it *under empty*."

This conversation never happened but it amuses me anyway, and helps me visualize what an actual vacuum is like. Besides, more humans than the French have been interested in pulling a vacuum out of thin air (ha, ha). Ancient Earth peoples, including, but not limited to, the Romans, had suction devices and pumps for various useful purposes, but basically didn't understand how they worked, and most interestingly hotly debated (as ancient philosophers were wont to do) whether when you pull water out of a tube, what remains is nothing (a vacuum) or just something kind of invisible (the nature of this latter part was even more hotly debated, because philosophy).

I guess I can see what they were all worried about. Everywhere you look, you see *stuff*. Some of that stuff is hard and rocky, some of it is light and airy. But no matter what, there's always something. It seems kind of sort of reasonable to state that, well, no matter what, there's always something. You can't have not-something. You open a window, and you can feel the wind rushing through—nature takes one look at an empty space and acts as quickly as it can to fill it.

In other words, nature abhors a vacuum.

As is usually the case in discussions like this, there were religious, philosophical, and vaguely mystical arguments back and forth on the subject.[1] When it came to the reality of the vacuum—the emptiest of empties—it wasn't just nature that hated a vacuum, but the divine. After all, a place with nothing in it would be truly devoid, which is tough to square with the concept of an all-knowing, all-seeing, all-*present* creator.

On the other hand, when you pull water out of a tube, there isn't exactly a lot in there, and seems kind of . . . empty.

Hence: heated debate.

Things got really intense in the 1600s when Otto von Guericke, who was supposed to be busy being the mayor of the German town of Magdeburg, instead idled his time playing with vacuums and pumps and suctions, because that's way more fun than being a mayor. He improved on some earlier designs and made a so-called vacuum pump, which could pull air out of a sealed vessel.

It was a pretty clever device. Attach one end of a hose to a chamber. Attach the other end of the hose to a piston. Pull the piston up, drawing air out of the chamber. Close a valve on the hose. Push the air out of the piston. Open the valve on the hose. Repeat as desired.

What remained in the chamber? Did air rush in to replace it in some sneaky way? Was there some mysterious, shadowy substance that could not be pulled out? Did the vacuum truly exist?

It's been a long time since the response to a vexing scientific question has been "let's get a bunch of horses," but von Guericke didn't mess around. He got a bunch of horses.[2]

Oh, and two hemispheres pressed together with the air between them sucked out with his fancy new vacuum pump. The bunch of horses couldn't pull the hemispheres apart, and these horses were strong.

They couldn't pull the hemispheres apart because there was nothing inside them, and a lot of stuff (air) outside them. The air was pressing in, and nothing was pressing out. The horses had to compete with all that pressure from the air, and just couldn't do it, because von Guericke's air pump was so good that there was a whole bunch of nothing inside those hemispheres.

In 1654, the vacuum won. We never stood a chance.

✪

So maybe the vacuum does at least, in a narrow sort of technical way that we can grudgingly allow, exist.

And then Isaac Newton went and did his whole universal gravity thing. Thanks, pal. In centuries past, folks thought that the universe beyond the Earth was full of all sorts of things. Mostly crystal spheres. That's right, crystal spheres. There was a crystal sphere for each of the heavenly bodies (one for the moon, one for the sun, one for each planet, and one for the stars). The spheres did the work of carrying said heavenly bodies in their regular route through the sky, with the Earth sitting serenely and boringly at the center (we didn't get a crystal sphere, I guess).

This was all fine and dandy until Johannes Kepler figured out that the planets in the solar system don't move in circles, but ellipses.[3] Oh, and the sun is at the center, not the Earth. It's kind of hard to have crystal, uh, ellipsoids that move effortlessly against each other through eternity, so the concept was just chucked like moldy leftovers.

Right, Newton. Nobody knew *how* the planets achieved this elliptical orbit business, until Isaac figured out that this "gravity" business isn't just an Earth business; gravity doesn't just pull apples from trees to the ground, but it connects every single object in the universe to every other object in the universe. Everybody creates and feels gravity simultaneously. The same gravity that pulls that apple from the tree is the exact same invisible force that keeps the planets looping around the sun in their fancy ellipses.

But here's the thing. If the Earth is hurtling around the sun at Great and Terrible Velocity, then what is it moving *through*? Nature abhors a vacuum, naturally and of course (except in limited cases where we have to work really hard, as in von Guericke's flashy horse-based demonstrations). But if gravity is doing all the work of keeping the planets in orbit, and the Earth and all the other planets are moving through something (whatever it is, but definitely not nothing), then shouldn't that be a real drag, man? As in, a literal drag? Something that makes it hard to orbit the sun? Something to slow us down?

But gravity could perfectly explain the motions of the planets, and there was no hint of any drag. No slow-down. No . . . thing.

Newton himself was officially Over It,[4] and was perfectly willing to toss the idea of stuff-between-the-stars in the scrapheap, but still, there was a lot of intellectual resistance to the concept of a great big vacuum. I know to the present mind, all informed and intelligent and aware of the universe that we are, it's hard to see what all the big fuss was about. So what if the universe is mostly void? What's the big deal?

I think it's fundamentally one of those concepts that is just so alien and alienating that it goes against our basic human intuition—the very concept that there can be *nothing at all*. But you and I were raised in a world of vacuum cleaners and outer space, so we're used to the idea. Let's give our intellectual forebears a break here; they had to figure out all this the hard way, so let's not be surprised if it took them a good century or two to sort things out.

You see, the story didn't end with Newton. The concept of space stuff resurfaced, again and again like the multiheaded hydra, appearing in various contexts and employed to solve various puzzling scientific riddles. It even acquired a name: the *aether*.

At first the ether (as it is more commonly spelled today) was proposed to be the stuff that planets swam through in their orbits, after the fall of the crystal sphere tyranny. It offered no resistance (otherwise the whole solar system would grind to a halt), it didn't affect light, it didn't taste like anything, and it basically didn't do much of anything else except sit there and *exist*, because remember folks, nature abhors a vacuum. But Newton rightly pointed out that the only reason folks wanted the ether to exist was because they wanted the ether to exist. If you were willing to abandon the desire for it, there was no need for it.

"Just let it go, man."—Newton (paraphrased)

But then there was this whole business about the nature of light. There are all sorts of bright things in the universe—the sun, the stars, fires, the usual—that emitted copious amounts of light. But what *is* light? What does it look and act like? How can we best describe it?

Newton and his intellectual cronies envisioned light to be made of teensy-tiny particles, called corpuscles (from the Latin for "teensy-tiny particle"). These little buggers blasted around the universe like a hailstorm of bullets, zipping from bright objects, reflecting and refracting, and eventually burrowing into our eyeballs to aggressively inform us of their presence.

To Newton, since light was just a particle, it didn't need an ether either—it shot around space just like the planets did, no big deal.

Then two very important things happened. One, Newton ended up dying, so he couldn't argue against anyone anymore. And two, Thomas Young conclusively demonstrated that light can instead act like a wave, at least when it wants to[5]. He showed this by shining light through two very narrow slits and duly observing on a screen behind the slits an interference pattern: in some places the light added together and got extra bright, and in some places the light cancelled itself out, leaving a dark strip.

This is 100 percent, precisely, exactly what waves do. Ergo: light is a wave. Take that, Isaac.

(Why nobody had thought to try this before Young in 1799 is anyone's guess.)

So sometimes light acts like a particle, and sometimes it acts like a wave. What's the big deal? The big deal is that if light is a wave it has to *actually wave something.* Think about it: a sound wave is a wave of air. A water wave is a wave of, well, water. When light is traveling from the sun to the Earth, what is the light waving?

This all got really serious a few decades later, when James Clerk Maxwell, a super-genius and also keeper of a truly unkempt beard, accidentally invented light.[6] He didn't mean to; he was just monkeying around with some equations trying to describe the behavior of electric and magnetic fields. But the equations neatly described how a changing electric field can create a magnetic field, and vice-versa. So it was possible to make changing electric and magnetic fields work off of each other, waving back and forth, capable of leap-frogging through space.

Curious, Maxwell took these equations and plugged in the known properties of electric and magnetic fields, and found that this leapfrogging electric-magnetic wave traveled at . . . the speed of light.

Light was indeed a wave; a wave of electricity and magnetism. But just like any other wave, it needed to move through something. There couldn't be true vacuum—there needed to be a (are you ready for this?) *luminiferous* ether—an ether that let light live in all its wavey glory.

✪

There was another, more subtle, more (shudder) *mathematical* reason for the ether to exist, and it had to do with viewpoints. Or, in the physics jargon: frames of reference.

In order to make the physics of Newton (gravity, action-reaction, forces and accelerations, all that jazz) work, there had to be an absolute frame of reference, somewhere out there in the universe. It didn't matter where it was or how it operated, but it just had to *be*. There needed to be some master ruler and master clock, allowing all of us to take measurements of position and velocity from that universal reference point. In Newton's math, all motion was relative with respect to that fixed frame.

But then came Maxwell and his beard to mess everything up (accidentally). In *his* electric-powered equations, the waves of electricity and magnetism (aka "light") had a single, fixed, constant speed no matter what. The speed of light was simply . . . the speed of light (yes, the speed of light can change as it passes through different substances, but that's a different discussion, you adorable nitpick you). The speed of light wasn't relative to this or that; no master clock or great cosmic timekeeper required.

This caused ulcers and elevated blood pressures across the globe.

Maxwell said that light doesn't necessarily need a frame of reference to go on and be light. But Newton said that all motion needs an absolute standard to measure itself against. Who wins? Hard to tell, so let's compromise: the ether. Not only was the (luminiferous) ether the weird ghostly substance that light traveled through, it was also the absolute reference frame of the universe. It was (or at least, resided in) the Absolute Stillness that was required to make Newton's laws work. It was within the ether that light obtained its fixed speed.

OK, not a bad idea. Let's go with it and see where it takes us. Let's perform an experiment!

If the ether is still and motionless, then our Earth as it orbits the sun is moving through it (even though we can't sense it in any way, shape, or form, but roll with me here). And if light can only achieve its constant speed in that same ether, then as the Earth swings back and forth in its orbit around the sun, then we must sometimes swim "with" the preferred speed of light and sometimes we must swim "against" the preferred speed of light.

Upshot: we should be able to measure changes in the speed of light.

In 1887, after years of failed attempts (Well, I hesitate to call any experiment "failed," because you always learn something, which is the point of experimentation, right?), the physicists Albert Michelson and Edward Morley were able to measure the differences in the speed of light using an interferometer: by splitting a single beam of light and sending those beams in different directions before recombining them, they could detect tiny changes in speed in one direction versus another.

The duo of M&M did the thing and found . . . nothing.[7] They couldn't measure any variation in the speed of light whatsoever. And if the speed of light was constant in all directions and at all times, then the need for the ether simply vanished into thin . . . uh, air.

The final nail in the coffin came from Einstein himself. While most people looked at the Maxwell versus Newton Battle of the Century and chose Newton (I mean, who wouldn't), Albert went rogue and declared

Maxwell the ultimate victor. But in order to defeat Newton, Einstein had to do away with the entire absolutely universal reference-frame game. Einstein claimed, and later backed up with some seriously intense mathematics, that *all* motion was relative. There was no such thing as a master clock and master ruler. You can only tell you're moving in reference to other things.

To Einstein, the speed of light was constant everywhere, at all times, for any observers. It just *was* the speed of light. The light invented by Maxwell didn't need an ether—it could move through the vacuum of space just fine, and it didn't need a fixed reference frame to have the speed that it had. In exchange, we learned that all motion is relative, as are all measures of time and space—time dilation, length contraction, $E=mc^2$, and all the rest of special relativity. If the absolute motion of Newton had to die, at least it went out in suitably grandiose fashion.

Within a few short years, the space outside the Earth's atmosphere went from being filled with a mysterious but necessary substance-but-not-a-substance to . . . just a vacuum.

In the early 20th century, almost two hundred and fifty years after Otto and his team of horses first conclusively demonstrated that nothing really does exist, outer space was born.

★

Before I get too far, I do want to be pedantic for a moment, because precision matters when talking about the great voids among the planets and stars. The vacuum of space isn't, usually, totally altogether empty.

I know, I know. Maybe I lied a little bit earlier. Just hold on, let me explain myself.

If I were to magically teleport you (and yes, that process would involve pure magic, but we'll get to that later) into some random spot in space, and you had at your disposal a detector capable of detecting the faintest, teensiest, most ethereal substances in the universe, you might be surprised to find yourself getting a few *ping-ping-pings* here and there.

Cosmic rays. Neutrinos. Radiation. Stray bits of molecules. Space fuzz, basically.

The universe is a busy place, and even the emptiest regions of the cosmos have *something* traveling through at any point in time. You simply can't avoid it. I mean, trust me, you'd have to have the most sensitive detectors imaginable to even catch a whiff of this stuff, but it's still there. Present. Most of it is innocuous enough, though we'll get to the deadly bits later on. It's empty *enough* for the likes of a massive object such as the Earth to fling itself through space, acting on the force of gravity alone, and not really notice. Hence why Newton and Co. didn't really notice.

I suppose it's a matter of pedantry to discuss whether outer space is truly empty, which is the entire point of this discussion. It's a matter of scale: if you have a big enough box and wait long enough, something is bound to swim through it. And should we count radiation, which is basically and annoyingly everywhere? Listen, I'll just leave it up to you. If you want to say that space is empty, I won't argue, because it's very much empty compared to the air you're breathing right now. But if you'd rather stake a claim on the not-empty side of the river, well then, you go right ahead—there's plenty of things to point to in the supposed nothingness of space and say, "Look, see, I told you there was something!"

But there's a very important detail regarding the vacuum of space I need to tell you about (besides the fact that it can kill you, quickly and horribly and enthusiastically, which we'll get to in a bit). If you were to take a random patch of space, say, that one right over *there*, and single-handedly remove every single particle, bit of radiation, dust grain; every neutrino, photon, neutron, electron; *all of it*, it wouldn't ever be quite, fully, matter-of-factly empty.

This is troubling to think about.

The vacuum of space itself is, in a vague but mildly accurate way, alive.

At the very least, it vibrates. It hums. It sings.

I know this sounds very hand-wavey and not scientific, so here comes the science. You're going to be spending a lot of time in the cold, hard vacuum of space in your travels, so you might as well have a crystal-clear picture of what's going on down there.

To understand what's going on, you have to realize that what we call a particle is nothing but, and basically everything else you've learned about physics is a lie.

In the modern, full-fledged picture of the cosmos, first developed in the mid–20th century by some physicists way smarter than you or me,[8] the fundamental particles of our universe are not single, solitary, lonesome little balls of mass and energy. Instead, they're pieces of something far larger and grander: a *field*.

Not the most awe-inspiring name, but we'll take it.

Every kind of particle (say, electrons, or photons, or top quarks, you get the idea) is associated with a special kind of *field*. This field permeates all of spacetime, extending across the domain of the universe since the big bang and onward into eternity, and from edge to edge (if the universe has an edge, but that's nothing you need to worry about right now). There's an electron field, a photon field (aka the electromagnetic field, you may have heard of it), a top quark field, you also get the idea here.

What we call a "particle" is, in the view of the branch of physics known as quantum field theory, a piece of a . . . well, quantum field. You take a patch of the universe, look for the quantum field you want to energize, energize it, and voilà: you get a bunch of particles hanging around that part of space. They can go on to live their little particle lives, and sometimes can even disappear—meaning that the field in that patch of the universe lost the energy to sustain their particley existence.

That's all weird but fine, but here's the real kicker: the quantum fields of our universe are never entirely quiet. It takes a mighty vibration to pinch off a particle from the field, but the energy of each field is never exactly zero. There's always a little bit of wiggling and humming going on, and sometimes those wiggles become randomly big enough to spit out a temporary particle.

That particle won't last long before the humming of the field reduces, pulling it back into the background. But still: particles are constantly popping in and out of existence, even when you don't want them to, even when you don't have enough energy present to make one yourself.

We call this background fizzing the *quantum foam* because that's an awesome and also descriptive name, and we call the nonzero background energy contained in the fields the *vacuum energy* because that's a boring but nevertheless descriptive name.

The vacuum of spacetime itself—with no other particles or radiation or anything else present—has an energy associated with it that you simply can never, ever get rid of. It's just *there*, an unceasing anxiety of the universe itself. You can't do any interesting things with the vacuum energy, unfortunately—all of our physics and science and life happen "on top of" that background energy, but it's there.

How much energy is there in the vacuum? Don't ask, because it's a doozy of an answer.

Shoot, you asked.

It's infinite. That's right: infinite. As far as we can tell (and we've been looking into this for quite a few decades now), the universe has an infinite amount of vacuum energy. Take a box, empty out that box (all the way). Congratulations, you now have a box with infinite energy inside of it.

It turns out that this infinity doesn't really matter, because like I said, all of physics happens on top of that energy, like living life at the top of a mountain. Who cares how tall the mountain is? What matters is the height of the chair when you fall off it while trying to hang a picture on the wall. The distance between your head and the ground is much more relevant to your impending trauma than the elevation of the mountaintop . . . even if the mountain is infinitely tall.

Confused? Good, welcome to quantum mechanics. I won't be bringing much quantum-this or quantum-that into my tales of warning and danger, for good reasons. One, we've got bigger fish to fry. And two, it's really, really mind-meltingly complicated stuff. My point here is that nature is full of surprises. Some of them benign, some of them nasty. What you may think of as pure vacuum is really a swirling, frothing, chaotic tangle of particles and radiation, created by everything from distant supernova explosions to the very nature of spacetime itself.

It's funny. Not ha ha funny, but "Hmmm, interesting, isn't it?" funny, that the story of the vacuum of space has come full circle. For centuries philosophers railed against the idea of the vacuum, for good reason. Then we discovered that it exists, then we discovered that, dang, most of the universe is made of nothing at all. And then it turns out that, well, nothing itself is, by very quantum definition, a thing.

So it goes.

But anyway, are you sure you want to leave your nice, warm, cozy atmosphere? Because all this madness is just the beginning.

✪

Look up. Go ahead, my little would-be wanderer. You dream of exploring the stars, right? Well then look up at them. They're far away—stupendously, frighteningly, too-far-to-think-about-in-any-sane-or-rational-way far away.

But the vacuum isn't.

Let's face it: The Earth ain't got much air. Oh sure, there are planets with a lot less of the gassy stuff, but this isn't a contest. If you were to blow up an egg to be the size of our home planet, then the shell would be thicker than our atmosphere.

Outer space isn't very outer. If you could drive your car straight up at a fast but efficient 60 mph, you'd experience the good, hard vacuum in less than an hour.

Can we all just take a moment to acknowledge, understand, and appreciate just how deeply *wrong* that feels? That everything you know and love, every breath of sweet, fresh air that you pull into your lungs, every microbe, every sound of laughter, every drop of water that you'll ever drink, exists within an impossibly thin bubble delicately wrapping an otherwise sterile, lifeless ball of molten rock?

I would say take a breath to relax, but that just might make it worse.

Atmospheres aren't easy to get, and very easy to lose—count yourself lucky you have a stable one.

Consider the Earth along with her nearest planetary neighbors, little red docile Mars and bright angry Venus. All three planets were born with an abundance of water, nitrogen, and carbon dioxide. They all came from the same protoplanetary ooze (NB the astrophysicists use other terms here); at these distances from the sun, the chemical and molecular makeup of the cloud of gas and dust that would eventually swirl and coalesce to form planets was pretty much the same. It was only through random chance that Venus and Earth ended up roughly the same size, with Mars putting in a good effort but coming up short.

Shortly after forming, the heavy elements (iron, silicon) sank deep down, while the lighter stuff (nitrogen, water) rose to the top. Presto: rocky planet with atmosphere. I'm skipping a few steps, but I'm sure you get the idea.

But these three worlds couldn't be more different, here and now, 4.5 billion years after their birth. Mars is cold, arid, lifeless; a frigid desert barely clinging to a status of anything more than just-another-rock-around-the-sun. Oh, it *technically* has an atmosphere, if you want to count a near-vacuum of almost entirely carbon dioxide as an atmosphere. Seriously, it's less than 1 percent the air pressure as over on good old Earth. On the other end of the spectrum we have Venus, choking itself on its tremendously thick clouds, again mostly of carbon dioxide (alright, nature, we get the idea, you like carbon and oxygen). It's so hot on the surface of Venus that it outcooks Mercury, despite being almost twice as far away from the sun.

Drop a bar of lead onto the surface of Venus. When you come back around in a few minutes, you'll just have a lead puddle. It's hot.

But if you could go back in time a few billion years ago (don't even bother trying), you would find three beautiful sister worlds, each sporting white puffy clouds, reasonably dense but not uncomfortable atmospheres, broad oceans, and beaches with plenty of great surfing opportunities. After all, the three planets all sit within what's called by serious astronomers the Habitable Zone (and cheeky astronomers the Goldilocks Zone) of our sun—where it's not too frigid to freeze all the water, and not too hot to boil it off in screaming agony.

Billions of years ago, the curious life-form would find three places it could potentially call home.[9] What went so wrong with those other two planets, and what went so right with Earth?

The problem with Mars is that it's frankly a bit too small. Not knocking you, Mars, it's how you were made, but unfortunately for you, how you were made didn't include the capacity to hold onto atmospheres and oceans for long.

The clue to the mystery of Mars's missing atmosphere lies deep in its core. It's solid, boring, lifeless core. Just sitting there like a big dumb ball of iron, not doing anything interesting. And *especially* not creating a strong magnetic field. In contrast, the Earth's core is molten, active, and spinning, whipping up a tremendously strong (OK, OK, it's a relatively weak magnetic field all cosmic things considered, but amongst the inner rocky worlds of the solar system it's a dynamic dynamo) magnetic field.

That magnetic field around the Earth is a *literal honest-to-goodness force field*, like something out of a cheesy sci-fi movie. It's invisible but completely envelopes the Earth, and is capable of deflecting away any incoming hazards like charged particles. Most of those charged particles come from the sun itself, and we give it the cute name of the solar wind, as if to pretend away its deadly intentions. I'll talk more about the solar wind later, but for now you just need to know that a) it's made of high-energy particles, and b) it's flooding the solar system.

Oh, and c) the charged particles of the solar wind can strip away a vulnerable planet's atmosphere in just a few hundred million years. As in, no time at all.

But thankfully the Earth can withstand this constant solar onslaught with its magnetic field: straight-up blocking some of the solar wind, and funneling what's left down into the polar regions, creating the beautiful light show of the aurorae. Next time to go see the Northern or Southern Lights, enjoy the spectacle for what it really is: that the Earth's invisible force field is keeping a lid on all that precious air.

Mars once had a thriving magnetic field, powered by a molten core that was similar to the Earth's. But because it's small, its core cooled down. And slowed. And stopped. And shut off the magnetic field.

Bye-bye atmosphere, it was nice knowing you.

With no more air the oceans evaporated. With no more water the place dried up.

Mars died as a youth.

And as for Venus? It suffered the fate of Icarus: it flew too close to the sun.

When our star was younger, it was slightly dimmer and slightly smaller: the dinosaurs of prehistoric Earth knew a weaker daylight than we do today (I'll get to the physics of the how and the disastrous consequences later, I promise). Long enough ago, Venus was nearly smack in the middle of the Habitable Zone, the region around a star where water can be as liquid as it wants to be, and oceans and atmospheres flourished in perfect harmony.

But then the sun got hotter.

And then Venus choked.

At first it wasn't much: the increased solar output drove up temperatures on Venus a tiny bit, which caused a tad more ocean water to evaporate and float around in the atmosphere. No big deal, right? Right, except for the fact that water vapor is a really awesome greenhouse gas, if what you're aiming for is a greenhouse effect: The vapor trapped extra heat on the surface.

That extra heat caused the oceans to evaporate juuuuuust a bit more, which put more vapor in the atmosphere, which increased temperatures, which evaporated the oceans, which increased temperatures, which . . . you get where this is going, I hope.

Water doesn't last too long in the upper reaches of an atmosphere before disassociating, but by then the damage had already been done. With no more water on the surface to lubricate its plates, tectonic activity on Venus slowed down and stopped (we're not sure today if Venus is totally 100 percent landlocked, but it's close enough for our purposes).

You would be gravely mistaken to think of a planetary atmosphere as totally separate and divided from its host planet. No, the two are intertwined; lovers, even. Carbon in the atmosphere can find itself inside rocks through various and sundry chemical processes, and plate tectonics can pull those rocks deep underground. Meanwhile, carbon can spew out from volcanos and other vents. The crust and atmosphere of a mild-mannered rocky planet act in concert to keep the balance of chemicals in check and ensure continued balminess.

But Venus went out of balance. With no more activity to pull carbon down into its mantle, the pernicious gas just kept piling up and up in the atmosphere. Carbon dioxide, another greenhouse gas, exploded to ridiculous levels in the Venusian atmosphere. With no water vapor left, there was no hope for the once-beautiful planet to regain its placid balance.

In the blink of a cosmic eye, Venus turned from a paradise into a hell, with no chance of recovering.

Sure, Venus is big enough to still host a molten core, but it spins so slowly (managing just two measly rotations every year), that it can't rachet up a protective force field. This means that its atmosphere is vulnerable to the same solar wind that stripped Mars of its glory, but it's just so dang thick it doesn't really matter. Venus choked itself to death once and for all.

In case you're wondering (and I know you are), yes, the sun will continue heating up, and yes, the exact same fate will befall dear old Earth. Take a breath of your sweet, sweet Earthly atmosphere while you still can.

✪

And if you're looking for a source of air on any of the other planets or moons in the solar system, I'll give you a giant heartfelt *good luck*. An inventory:

- Mercury. While this little world *technically* has an atmosphere, it's really straining the definitions of both the words "technically" and "atmosphere." We can only detect it with our most sensitive probes. Good luck trying to breathe it, or just generally survive on Mercury in general. Either you'll be roasted alive with the full force of the up-close-and-personal sun, or immediately locked in ice and darkness. With no air or water to circulate heat from the warm side to the cold side, it's just that strictly binary. Either way, nothing to pull into your lungs.

- Venus. We talked about Venus. Chokingly thick. Can literally melt lead. Sure, as you go higher up in the atmosphere, the temperatures and densities drop, until at a certain high altitude the surrounding air is something vaguely resembling room temperature and sea-level pressure. But don't expect a nice salty breeze. No, one of the main players in the dystopian Venusian atmosphere is sulfuric acid. You know, acid rain. An atmosphere *literally made of acid rain*. Have fun!

- Earth. White fluffy clouds. Gentle breezes. A paradise. Treat her nice.

- Mars. Come on, you call that an atmosphere? Yes, you do, but you shouldn't. And it's carbon dioxide. Unless you're a plant this isn't going to be any fun.

- Jupiter. Oh yeah, loads and loads of atmosphere. So much so that if you were to dive in, by the time you got a few miles down you would be crushed into oblivion like a uniquely squishy soda can. And even higher up, you have to contend with winds racing at hundreds of miles per hour, hurricanes galore, and one massive storm—called the Great Red Spot because it's red, it's a spot, and it's great—that could engulf the entire planet Earth two times over, and it's been raging for hundreds of years *at least*.

- Saturn. Like Jupiter, but less so. Did you know that there's a hexagon on Saturn? Yeah, at the north pole there sits a massive cyclone, and a retinue of storms at lower latitude circulate high-velocity winds in the shape of a hexagon. Just think about this for a moment: a hexagon. On Saturn. Do you even want to go near something that weird?
- Uranus and Neptune. Nobody really talks about these two ice giants. They do have super thick atmospheres made of hydrogen, helium, methane, and trace amounts of "other." Look it's just really cold out here so take a hard pass.
- Pluto. Featuring a wispy atmosphere of mostly nitrogen (just like the Earth!), it occasionally snows here. So yes, you can go skiing down the slopes of the Plutonian mountains with some fresh powder. But bring an air tank and a few layers.
- Titan. Now this massive moon of Saturn is intriguing. It has one of the thickest atmospheres in the whole solar system, completely obscuring its surface from remote view. But while it's a thick and hazy mush of nitrogen, methane, and hydrogen, it's beyond frigid, sitting at less than 100 kelvin (K).

That's it. Seriously, that's it when it comes to atmospheres (and sometimes even that's straining the definition) in the solar system. You want breathable air? Stick to Earth or bring it with you. And once you're outside the system altogether . . . well, *be prepared*.

★

By agreement amongst people who agree upon such things, once you're about a hundred miles up from the surface of the Earth, you're in space.[10] That rough line comes from the observation that at that elevation, the air is too thin to support lift on wings. In other words, a hundred miles up, you can't use airplanes to get around anymore; it's rockets or nothing.

There's still *some* air, annoyingly enough. If you're in orbit around the Earth at this distance, you'll have to worry about atmospheric drag, slowly but steadily sucking energy away from your orbit. If you don't give yourself a little boost every once in a while, you'll find yourself becoming a meteoric lightshow for anybody along your ill-fated path to ground level.

Air may be a relatively thin substance, but it can pack a wallop when you're slamming through it at tens of thousands of miles per hour.

But even though the air up past 100 miles is still thick enough to mess with our orbiting machinery, don't bother trying to breath it, because you'll die in a very gross way.

Should you happen to be exposed to the vacuum of space without a suit or source of oxygen, things go haywire fast.

The first thing you're likely to notice is a quick nasty flash-freeze as all the oils and liquids on your skin (like the sweat that built up as you contemplated what was about to happen) immediately crystalizes and evaporates. Air pressure literally keeps liquids in their place on surfaces, just like the Martian atmosphere did for the oceans on that now-tired and sad planet. The instant formation of ice crystals on your skin, your eyeballs, your armpits, and probably even under your toenails will cause quite a bit of pain.

Needless to say, the surface of your skin will become bone dry. You're going to need a serious slathering of body lotion when and if you make it back to safety.

But, nasty as it is at first, you're still alive. Heart beating, brain thinking, limbs flailing. Helpless and drifting, but alive.

Some old sci-fi movies loved to show the effects of getting "spaced" (and what a wonderful euphemism that is). You might blow up in the vacuum, your eyes might bug out, you might instantly freeze to death.

None of this happens. Oh, you'll still die quickly, but it's from something else.

You won't explode, but you won't exactly be comfortable, either. The insides of you are full of fluids and gases and delicate tissues, and in a vacuum the outside of you is full of absolutely nothing. In that environment the fluids and gases and delicate tissues inside of you would really love to expand and explore their newfound larger territory (i.e., "explode" in the common parlance) but unfortunately for them, your skin is really good at keeping all your insides on the inside of you.

Try as they might, the tension provided by your skin is more than enough to keep you intact against the vacuum of space. Phew.

But wait, there's more! Fluids and gases *near* the surface of your skin will still *try* to get outside of you, which is bad for you, but instead will have to be content with just simple swelling and expanding, pushing your skin to its limits. Still bad for you, but slightly less so.

This swelling under a vacuum is known as *ebullism*, not to be confused with an embolism (which is also deadly but for decidedly non-vacuum-related reasons). How big you'll get isn't exactly known, since nobody has been audacious/unethical enough yet to dump a live human into a vacuum chamber to see what happens. However, there have been a few mishaps and near-misses in the history of human spaceflight, so our current best guess is that you'll plump up to about twice your current volume once you hit zero on the pressure gauge.

Still, should you be recovered, you'll quickly deflate to your normal self (kind of) and eventually heal.

So the vacuum of space won't kill you directly, it will just make you supremely uncomfortable.

What gets you is the oxygen, or rather lack thereof.

You see, in the vacuum of space there's no air (kind of the point), which means your lungs are empty. But your heart hasn't clued in yet, it just dumbly beats-beats-beats as if everything were normal (and might even be beating a bit faster than average, considering the situation you just found yourself in). The job of the heart is to carry blood up by the lungs where it can grab some of that precious oxygen, then proceed to drive that oxygen-rich blood to all the parts of your body that need it, which is all of them.

So the blood train pulls up at the lung station to grab a load of O2, but the warehouse is empty. But with a toot of the whistle and a chug of the engine, off the train goes into the arterial sunset, its cars empty.

Within a few seconds, your brain, the most oxygen-hungry part of your body and a somewhat critical organ, notices that it's not getting its regular supply. It automatically goes into power-saving mode, turning off less-than-critical functions like consciousness.

In less than ten seconds after exposure to the vacuum, you take a nap.

Now, you're not dead. Yet. If you're pulled to safety you can still wake up and go on to live a comparatively stress-free life. But as the seconds turn into minutes, the rest of your organs also take note of the lack of oxygen and begin to shut down, one by one. Eventually, all your organs shut down.

This process is known to the medical communities as "death," and is generally advised to be avoided.

The vacuum of space will kill you—all the way dead—in about two minutes, your bloated, skin-dried corpse doomed to float through the vastness of space forever.

You might be tempted, when faced with this grim fate, to hold your breath, to save one last pull of oxygen to give you just a few more seconds to clamber your way to the safety of a pressurized and oxygenated vessel.

I have only one word in response to this suggestion: don't.

The air inside your lungs is held at a nice and familiar level of one unit of atmospheric pressure. It's the same pressure as the air that was outside your lungs when you took your last breath. The pressure outside your lungs in vacuum is zero, because it's a vacuum and that's the definition.

I'll cut straight to the point: the muscles and slimy tissues of your throat were not designed to hold one atmospheric pressure of air against a vacuum. That air will make its way out, despite your best efforts, and it will do so quickly and violently, expanding into the vacuum as it does, damaging (perhaps permanently) your delicate throat and damaging (definitely permanently) your even more delicate alveoli, the little tiny sacs inside your lungs that do the job of delivering oxygen to your now-deprived blood.

Again, don't. Just let that air go. It's not worth the pain and misery. If you try to hold your breath and you are somehow miraculously recovered, then it will take yet another miracle to actually stitch you up.

Once you hit vacuum, the clock starts ticking, and you better think fast.

At least you won't freeze to death. You'll be long gone before that unique misery sets in. Yes, space is cold. It's hard to strictly assign a temperature to the vacuum of space, because there's hardly anything (and sometimes even nothing) in there, and you need lots of microscopic things whizzing about to define a temperature. For various technical reasons of no immediate concern to your survival, we assign a temperature of 3 degrees above absolute zero to space, but for now let's just call it "cold" and move on with our (hopefully long) lives.

Or rather, the end of them. Your body is warm, and space is not. Therefore, heat will flow from your body and into space, making you, eventually and regretfully, cold. And there are three ways of moving heat from one place to another: conduction (touch something cold, and the jiggling of your atoms will start making the cold atoms jiggle too, transferring energy and heat), convection (a fluid like air or water circulating around you, again with the jiggling of atoms to transfer heat, but the motion of the fluid continually replenishes with a fresh supply of new atoms ready for the jiggle), and radiation (your atoms give off light, which shoots out of you and hits some other random atom, making it jiggle).

Lots of jiggling involved in the flow of heat.

In space there's nothing to jiggle. You're not touching anything, and there's no air or water to flow around you. Conduction and convection are dramatically good methods for removing heat, which is why hypothermia is such a big deal even in tepid water.

All that's left in space is radiation. A typical human body radiates energy at about 100 watts, the same as a reasonably bright old-school incandescent light bulb. You can't see the radiation that humans give off, because it's mainly in the infrared portion of the electromagnetic spectrum, unless you have infrared goggle. Then we light up like the limbed bulbs that we are.

We normally have all sorts of mechanisms of generating heat to counter that continued loss from radiation. We generate heat from the food we eat, from things we touch, from the sun itself, and just generally from our environment. But still, we have to work pretty hard to keep our body temperatures in the happy zone.

In space we don't get that luxury, especially when we're dead. We're just warm lumps of meat and water, at a temperature of 98 degrees Fahrenheit (F), emitting 100 W of radiation into the void, which takes a surprisingly long time to cool that lump down.

If you're in the full blast of our sun's light, you might not even freeze at all, depending on your particular orbit. If you're near the Earth, remember that the amount of sunlight that our home planet gets is just the right level to keep liquid water nice and liquid. Which is a lot of energy. So if you need to squint your eyes (let's ignore the damage done by your tears freezing) to see in the vacuum of space, you can rest (permanently) assured that you won't turn into a meat popsicle.

You're going to get a really nasty sunburn, but that might or might not be a major concern of yours, depending on how those precious few minutes pan out.

But if you're too far from the sun or stuck in some shadow, you're iced. Within the span of a few hours, your body will slowly, slowly, slowly lose heat, dwindling down to the same near-absolute-zero temperature of your surroundings, the water in you turning into ice crystals, locking your limbs in whatever posture you wish to give as your final message to the universe.

Over the course of eons, micrometeorites—tiny little specks of dust that we will explore in haunting detail later—with carve tiny but noticeable chunks in your now-frozen flesh. Given enough time (and the universe has plenty of that to spare, I assure you), I suppose you'll eventually dissolve from the continued microbombardments, with your constituent molecules eventually spreading out amongst the stars, perhaps—one day long distant from now—to join in the formation of a new solar system.

Technically, I suppose, you will manage to travel among the stars.

At least, parts of you.

Asteroids and Comets

Rocks and ices.
The tumbling leftovers.
Too many to count
Endless ways to ruin a good day.
                    —Rhyme of the Ancient Astronomer

Let's start with some good old-fashioned dangers. Not body-bending black holes, not skin-melting exotic explosions, not brain-numbing relics from the ancient universe (though don't worry, we'll get to all that). For now, rocks. Just rocks. Rocks with attitude. Rocks that are looking for a target and, whoops, you've got a bullseye painted right on the side of your ship.

Lots of rocks. Little teensy microscopic grains of dust. Pebbles the size of . . . pebbles. Boulders with enough gravity to form miniature ring systems. Hulking masses just shy of being called planets, but don't quite get to be classified as one. Don't tell them that, though, because they have a short temper and are surprisingly agile for their size.

Comets, asteroids, meteoroids. Apollos, Trojans, Trans-Neptunian Objects. Whatever you call them, however you classify them, they're trouble. They can punch a hole in you and your ship before you can blink. They can pulverize you so completely that another explorer a hundred years later will think that you're just more space dust.

They're dark, quiet, and *fast*.

I would tell you to just avoid them altogether, but they're *everywhere*.

You think you're smart? You think you can outrun them, shoot them before they shoot you, detect them, avoid them? You know what they say: the bigger they are, the more trouble they attract.

Space joke. Never mind.

Let's break it down before you get broken down.

First you have your generic space dirt. Just microscopic bits of carbon, silicon; maybe some water or minerals. Nothing too special, often chained together. They're the leftover debris from when asteroids, comets, or space-ships collide. They're found all over the place, too: not just the solar system, but in the vast distances between the stars. They get pushed out there by the sun's radiation only if they're tiny enough; otherwise they slowly spiral in to be burned to ash in the sun. Or rain down onto a planet. If you're on such a planet right now, there's a steady stream of this gunk drifting on to you as you read this. Gross.

It's a little bit hard to estimate (because, small), but we think somewhere around a few hundred tons of random dirt is gently drifting onto the Earth every single day.[1]

At the small end they're called simply "dust" (I know, how original) and on the big end they're known as micrometeoroids, with the strict technical scientific definition of being "tiny."

You can actually see this stuff. Not individually, of course, but spread out as they are across the solar system, they reflect a lot of sunlight. If you get yourself out on a dark, clear night, find the plane of the solar system (easiest if you have a couple planets in view, then just play connect-the-dots with them), and you should be able to see a faint glow starting at the horizon of the setting sun and tapering upward. It's called zodiacal light, and it's the accumulated reflection from trillions and trillions of tiny space bits in our solar system.

Actually, a bit pretty, if I do say so myself.

But don't let the prettiness of the zodiacal light fool you. Micrometeorites are No Fun. Despite their diminutive size, they are still traveling at typical speed of over 20,000 mph. Have you ever been hit by something traveling at 20,000 mph? No, you haven't, because you're still reading this book.

Protection against micrometeoroid impacts was one of the first challenges that humanity had to face when we upgraded "Earthsuits" to become "spacesuits" and the earliest attempts were simple and straightforward: wrap our astronauts and cosmonauts in thick layers of padded material, and cross our fingers that any tiny zipping rocks just get buried in there, rather than something more fleshy and delicate.[2]

Next up are meteoroids. Still pretty small, but big enough that if you saw one you would say, "Hey, there's a meteoroid. Right over there." Definitely larger than dust grains, but generally considered to be no wider than a meter or so.

They too are smashed up versions of their larger rocky cousins orbiting around the solar system, or chunks of larger objects that just happened to fall off because that happens sometimes. They're too small to have enough gravity to make them round (remember, gravity loves to pull everything *in*, so that force will do anything it can to remove lumpiness, but there's only so much effort it can put into the job if there isn't a lot of mass to go around), so they're in all sorts of jagged, ragged shapes. They tumble along in their orbits minding their own business, unless you get in their way. Then they'll rip through you like soft cheese.

Think they're harmless because they're so small and cute? See, they're hard—and now they're getting big enough that they can have a significant proportion of metals inside them. And when they strike you or your planetary atmosphere, they're usually moving at about 20 kilometers per second, which is about 45,000 miles per hour.

Or, in more familiar and scary terms, around fifty times faster than a bullet. A bullet made of solid rock, a foot or two across. And composed of dirt, so it isn't exactly bright and broadcasting its location. On the Danger Scale, these are a solid eight out of ten.

Occasionally one of these meteoroids will fall onto a planet. Ever see a really fantastic shooting star? One that lights a flaming trail across half the sky? One that you can make a wish on? One worthy of a good set of "oohs" and "ahs?"

It was made by a meteoroid about the size of a grain of sand, careening into the planet's atmosphere at about 100,000 mph. At those extreme velocities, the rock literally pushes on the air in front of it like a piston, squeezing it down and heating it up to temperatures ridiculous enough to a) turn the air into a plasma, and b) vaporize the rock, bit by bit. Hence the nice bright glow and the fiery tail as the meteoroid plunges to its doom.

Because ancient Earth peoples thought that these light shows were some sort of atmospheric phenomena (and I guess they technically are, because you only get the sizzle when the rock hits the atmosphere), they attached their generic word for weather-related happenings to it: hence, *meteors*.

If the meteor survives the trip and makes it to the Earth's surface, it gets a new name: a meteor*ite*, and surface-dwellers can collect these rocks and sell them on eBay.

While most of the 25 million micro to regular meteors encountering a planet like the Earth every day evaporate in the upper reaches of the atmosphere, there's only one word to describe the ones that finish that hundred-mile journey downward: *boom*.

And as for the even larger rocks. Yikes. These bad boys can be anywhere from a few feet across (i.e., just on the line of meteoroid) to might-as-well-call-them-planet-sized (and that's always up for debate amongst our astronomical friends). If a larger body is in a known stable orbit, is made of mostly rocks and/or metals, and don't bother anybody, we call them *asteroids*. If they're in unstable orbits, live most their lives in the outer solar system, are made of mostly ice, and cop an attitude, we call them *comets*.

If one of these big suckers smacks a planet it's just plain bad news. You don't get to watch and enjoy the fireworks. Instead you run for your life and start worrying about where your next meal will come from.

✪

Almost all these rocks of various shapes and sizes are the leftovers from when the solar system first cooked up. You ever stir some flour in a bowl, and get a little too excited or the dog starts barking or something interesting finally happens on that TV show, and flour goes all over the counter? It's a little like that.[3]

Young solar systems are messy places. We'll talk later about why you should avoid them in general, but one of the reasons they're so dangerous is all the dang rocks flying around. Big ones, medium ones, little ones. Planets need to form, and to get a planet from a cloud of gas, you need a lot of bumping and rubbing. Dust grains turning into rocks. Rocks gluing together into rubble piles. Rubble piles attracting their neighbors to form protoplanets. Protoplanets crashing into each other to form . . . well, planets. The bigger the protoplanets get, the more stuff they pull onto themselves with gravity, and the more encounters they have with each other.

But it's not nice, neat, and orderly. It's not a steady progression of pebble-sized Earths to proto-Earths to regular-Earths to Earth-Earths. There can be multiple planets forming within or near the same orbit. Two protoplanets enter; one protoplanet leaves. There were dozens, if not hundreds, of protoplanets swimming through the early solar system. Today? Just eight (or nine, depending on who you ask, but we're not stepping into *that*). To make that happen there had to be quite a bit of violence, and that violence was not uniform—every collision left behind a floating rubble pile, a weightless reminder of the devastation that occurred.

We even think that the Earth's own moon formed from such a collision of protoplanets. Think about it: of all the inner planets, the Earth is the only one with a major moon. Mercury: none. Venus: none. Mars: two, but they're dinky. The only way we know of to get a decent moon around the Earth is to have the young planet experience a cosmic car accident with a Mars-sized object. That collision had enough raw *oomph* to send a moon-sized blob into orbit around the young Earth.

Now imagine that happening to you.

After this phase, an early planet might think it's won the competition, cleared out its neighborhood, is all nice and safe, and can get started making an atmosphere, oceans, and little bugs to start crawling around on its surface. But before you know it, a Jupiter-sized gas giant decides to have a visit to the inner solar system to see what it's like.

"Gas giant in the inner solar system" should be the phrase to replace "bull in a china shop."

These migrations happen to some solar systems—luckily not to ours, otherwise all the inner planets would have been scattered either to be swallowed by the sun or flung outside the system altogether. But our own home wasn't without a little do-si-do of its own, but as chance would have it most of the mega-engineering rearrangement in this system happened in orbits well away from the sun, resulting in a delightful little event known as the *late heavy bombardment*.[4]

Oh, the bombardments. When a protoplanet attracts a nice big boulder to add to its collection, it's not like they just hug and get along forever. No, there's a crash, a bang, and a boom. And during this age, comets and asteroids from the outer solar system, newly destabilized from the cosmic reshuffling, rained onto the inner worlds, striking blast after painful blast, each slam leading to a new round of meteors ejected out into interplanetary space.

All this activity, all this commotion, makes quite a big mess. Sure some of the fragments will eventually find their way back down onto one of the major planets, but a good fraction—good enough to give you a headache—will remain spread throughout the solar system. Some of the bits stay in a stable orbit and can hang out for billions of years. Some of the bits get flung beyond the outer planets and almost into interstellar space. And some just go nuts.

Heck, we've even found samples of Earth dust on the moon, and Martian rocks on the Earth. Let's pause for just a moment to consider the raw energies required to deliver one piece of a planet to the surface of another.

Yeah, wow.

The bits of debris, whether leftovers from the primordial era of our solar system or blasted into space in more recent epochs, which stay near the sun get their water and carbon dioxide blown away by the heat, leaving just the hard rocks. Anything fragile gets ghosted into space dust. Only in the outer reaches can the ices and the weaker materials stay together, clinging to each other in the lonely dark.

The furthest bits may not even have had a chance to form something bigger at all. They may be the leftovers from the formation of our home solar system, mixed with the leftovers from other systems in the neighborhood, forever dithering between one system and another, never having the opportunity to join the party. The gravitational pull from one system was never strong enough to overwhelm the others and bring it home. They snooze, they lose.

But that's not all. Back in 2017 on Earth, astronomers spotted something quite odd in their usual searchings of the heavens: a bit of space rock. That itself was nothing new, and assumed to be just another comet inbound from the outer solar system. But continued tracking revealed a few unfamiliar facts. For one, it was moving blisteringly fast, and I mean *blisteringly*, with a maximum speed of over 87 kilometers every single second, or close to 200,000 mph (and you can see why astronomers and explorers tend to switch units when dealing with anything off the surface of the Earth).

That's faster than the speed you need to escape the sun's gravitational well. In other words, this mysterious rock was on an escape trajectory.

What's more, it was coming in at a very odd angle, far more inclined than anything (literally, *anything*) else in the solar system, almost perpendicular to us.

With that speed and that orbit, there was only one conclusion: humans had just witnessed their first interstellar interloper.[5]

Named, 'Oumuamua (a Hawaiian term that roughly translates to "scout"), by the time it was tracked and confirmed it was already on its outward journey from the sun. Follow-up observations could barely see it; it was small, only a few hundred meters long, but extended like a cigar. It had a deep red coloring, similar to other objects found in the icy depths of the solar system. But it had no trace of the ices that make comets comets; no tail, no visible outgassing.

Just a strange, alien rock. Tumbling end over end, having traveled for tens of thousands of years to reach us from some unknown origin, and headed outward for another sojourn into the great expanse for another collection of eons. Whatever sent this shard hurtling into interstellar space is almost too frightening to contemplate.

Rough estimates pin the number of large foreign intruders (The "large" specifier is needed because there's always some random space-bits floating in from beyond the reaches of the sun, and should we really count those?) at around one per year. But in general they're small, dimly colored, fast moving, and altogether hard to spot.

Just how many have passed by you without you ever noticing?

✪

Let's sketch out the most dangerous parts of the solar system:

The solar system.

There, that was easy.

Oh, you want *details*? Fine. I'll use the solar system as an example, but of course other systems will have their own quirks, so be on the lookout for each new system you enter, especially if they're unfamiliar or uncharted. There will be, however, a few danger zones common to pretty much any system, since many of these regions are byproducts of planet formation. These will give you a good place to start—or rather, a good place to stop.

The tiny micrometeoroid cosmic dust (i.e., the stuff the solar system should've tidied up ages ago if it expected company over) is spread pretty thinly across the entire system. Individual grains are generated wherever there are more space rocks than average, leading to little collisions here and there that chuck off the bits of nameless debris.

Each individual grain is either drifting inward to the sun or gently surfing out, depending on how heavy it is—naturally it's a little bit hard to predict, since each tiny bit is subject to its own peculiar set of forces. New dust is always being created; otherwise either the solar pressure from the sun or the gravity from that same sun would've blown away or gobbled up all the specks, leaving everything neat and clean. All it would take would be a few million years of persistent housecleaning, but since the solar system is much, much older than a few million years, *something* is still out there, making a mess of the place.

Go outside the plane of the solar system and the dust drops off almost immediately, since all the dust-making rocks largely live amongst the planets.

In general, you don't have to worry too much about the dust, assuming you are properly protected. And if you aren't, what are you doing out here?

Besides the ever-present and always-aggravating dust, the nearest and perhaps most dangerous hazard zone in the solar system is the Main Belt: a great loop of asteroids and meteoroids between the orbits of Mars and Jupiter. Here you'll find Ceres, the largest of the asteroids. It's so large that, as new generations of astronomers rebel against their predecessors, it occasionally gets called a planet. Or dwarf planet. Look, just don't ask too many questions.

The fact that asteroids like Ceres and Eros are called "Ceres" and "Eros" instead of "HKJ-33028472" and "RGF-92750285" is a clue of just how big these things are. Not everybody in the solar system gets a name, after all.

This also means that the Main Belters are relatively well-known and well-mapped. I mean, come on, if dudes in the 1800s could spot them, then you ought to know where they are by now.

But even with a size a few hundred kilometers across, a good strong jump could launch you into escape velocity from their surfaces.

Of course there are a lot of smaller asteroids that don't get named after mythological creatures. You can always name them yourself—I hear Gertrude's Dream and Hanklepix 1776 aren't taken—but don't expect anybody else to recognize your classification. That only works if you're the first to discover something, and trust me, it's unlikely you'll be the first to discover a Main Belt asteroid.

Astronomers used to think that the Main Belters are the leftovers of a gigantic planetary collision, because those were violent times and the astronomers didn't know any better. But if you add up all the rocks in the belt, including the biggies like Ceres, you'd end up with the wimpiest, puniest, lamest planet in the inner system, not even half a Moon.

Instead you need to blame Jupiter. Why not, it's not like Jupiter cares. It's too big to care. And it's big enough to prevent anything in the Belt from being big: usually one planet doesn't affect the birth or life of another. When was the last time you worried about Venus or Saturn? Once a planet clears out its orbit, it's like, "Don't worry guys, I got this. This is my space. I won't bother you and you won't bother me, you dig?" Everyone stakes out their turf and that's that.

Except for Jupiter. The Big Bully. A planet may have *tried* to form at the orbit of the Belt, but Jupiter had better ideas. In the Belt, Jupiter is big enough and close enough to give an occasional gravitational tug on the asteroids. Two asteroids decide to get closer and become something bigger? Nope, says Jupiter, not this time, and pulls them apart. A larger asteroid begins to vacuum up its neighbors in the early solar system? Think twice, says Jupiter, and ejects all the material out of the belt.

What a jerk, but we don't call Jupiter the King of the Planets because he's a monarchal figurehead in a representative democracy.

So we're the left with the Belt: the planet that never was.[6]

I know you're thinking that the Belt must be a tricky place: asteroids careening off each other, smashing and crashing, a continual frenzy of shrapnel and danger. You're right, but not in the way you think are. Asteroids do crash into each other. Shrapnel—aka meteoroids—does fly around randomly and unexpectedly. The Belt is one of the main sources of cosmic dust in the system.

But I need you to think slower.

No, not as in *think more slowly* but as in *the things you're thinking about are slower.* All that craziness happens for sure, just on slower timescales. A Main Belt asteroid may go millions of years before bumping into a neighbor. Ceres has craters, but those were made long ago. From the perspective of the solar system, with its billions of years of history, the Belt is a crazy hot mess of a place. But from a human perspective, it's barely more active than Aunt Maude during one of her afternoon naps.

There are more bits of jagged, horrible rocks in the Belt than average, but when your "average" is pretty much zero, it's not exactly a gold medal achievement. If a typical asteroid is about 10 meters across (a random number I plucked out of my head), then the average distance between asteroids in the Main Belt is so large that you would need to string about 100 million asteroids end to end to get to your next nearest neighbor. I hope you can see how we've been sending craft through the Belt for years with nary a scratch or scrape.

Once you've plotted out the locations of the biggest ones—and since these have been known for hundreds of years, it shouldn't be a problem, even for you—you just have to aim your ship and cross your fingers. The meteoroids are too small to keep track of, and new ones are constantly being created. The Belt is so *big* that at any one time two asteroids are probably hitting each other, somewhere.

In a spaceship, you have two choices if you're traveling to the outer system: after passing Mars, you can either jump up out of the plane of the solar system, avoiding the Belt entirely, or just plow ahead. Jumping up out of the solar system is easy: you can use the momentum of Mars to slingshot you out. But getting back in is a bit trickier, since there's nothing up there to help you get back down.

So, plow ahead it is.

Basically the Belt represents a slightly elevated risk of getting blown to bits. Not big enough to make us stop sending craft hurtling into the outer system, but it should be big enough to make you think twice. Will you be the unlucky traveler to meet the Big End before you've had the chance to forge your dreams among the stars? Will the one-in-a-million chance of getting personal with an unknown meteoroid traveling at tens of thousands of miles an hour be your unlucky number?

Every day is a roll of the dice, and to get to deep space you've got to play the game.

✪

I don't want you to think that the Main Belt is the only source of shorter-than-average life expectancy. The Belt is rather stable nowadays. Sure, it had a turbulent youth—who doesn't? But over the eons it's settled down into predictable orbits, only occasionally reliving its past and attempting something dramatic.

It's beyond the giant planets where things get wild.

Someday they'll tell tales of the Taming of the Outer System, of how we turned a wild region into a thriving civilization, free from struggle and violence. But that day is not this day. No, today the region past Neptune, in the last stretches of the solar system, where the light from the sun is barely a glimmer, is full of nastiness and surprises.

It's a dark and frozen space, full of icy threats.

Astronomers and explorers are still busy charting this region of the solar system, which is incredibly difficult because everything out here is a) very small, and b) very far away. In general, every object past the orbit of Neptune is known as a trans-Neptunian object (TNO), because of course it is.

By far the most famous TNO is Pluto, an object so big it was once considered to be a planet, then reclassified as a dwarf planet, but still a contender in some circles for the big title.[7] It hosts a (relatively) giant moon, Charon, but parts of its surface are surprisingly smooth. Apparently, according to our latest probes, Pluto has a gaping, festering wound in the form of a giant nitrogen ice glacial field, ringed by water ice mountains the size of Mount Everest back on Earth.

Something is managing to keep Pluto warm enough (out here the sun is just an angry pinprick of light, locking these worlds in permanent twilight), but to date we don't understand what. Did I mention that Pluto might also be hiding a liquid water ocean underneath its icy shell? Yeah, weird stuff out here, for sure.

But Pluto isn't alone. It has four other tiny moons. Then there's Eris with its moon Dysnomia. Strangely elongated Haumea. Makemake. Quaoar. Sedna. Orcus. Salacia.

Given these incredible distances from the sun, these worlds are absolutely massive. While some are gray-blue from the mixture of ices on their crusts, some of them have the dull-red ochre found only in the outer reaches of the system; organic molecules spoiled by too much UV radiation built up over eons.

With the exception of Eris, these worlds and countless smaller ones belong to the first ring of danger in the outer system, the Kuiper Belt. Like the Main Belt, the Kuiper is filled with rocks of all sizes. Sure, one or two of these might be worth a quick visit, but if you've gotten this far, then you're on a trajectory and speed that will take you clean out of the solar system, no time for chit-chats with lonely cold balls of rock.

Also like the Main Belt, the Kuiper is relatively calm and stable. "Stable" has a very precise scientific meaning, which is important for travelers like yourself. If you're sitting on a chair and someone bumps you, you may wobble a bit but otherwise stay where you are. That's *stable*. If you're standing on your tiptoes at the top of a mountain peak, and someone bumps you, you may be in for an interesting descent. That's *unstable*.

The Main and Kuiper Belt objects are pretty much stable: they may get a little gravitational nudge here and there, but for the most part they stick where they are in known orbits. Every once in a while a few will escape, sometimes to become a moon of one of the gas giants, but otherwise they're trapped in their lonely, icy prison. Navigating the Kuiper is the same as navigating the Main: make sure you're not aimed directly for one of the major players, and hit the throttle.

But past the Kuiper Belt, starting about 50 AU (Have I mentioned AU yet? Astronomical Unit. The average distance between the sun and Earth. No points awarded for guessing which planet's astronomers named it.) to more than 100 AU is a much more dangerous region: the Scattered Disk. Its name even sounds scary. The Disk is unstable, unruly, and unrepentant.

It sows discord and chaos throughout the rest of the system. Were it not for the Disk, the inner solar system would be a much safer neighborhood.[8]

The Scattered Disk is one of the great homes of the comets.

Left to themselves, comets pose no more threat than their asteroid cousins. Just slightly icier versions, since they formed far away from the sun and didn't get their water boiled off. But in the Scattered Disk, life isn't pretty. You may be in one orbit, and before you blink you've been pushed into another. And another. And another. And so on until you find yourself either ejected from the solar system completely or headed straight into the burning fireball at the center of it all.

All that nudging, bumping, and general scattering is due to the giant planets that orbit inward of the Disk. As those massive gas and ice balls make their lazy way around the sun, they tug and tweak on the objects out here. After enough persuasion, they destabilize from their cozy and familiar orbits, and just like falling off the top of a hill can send you in either direction, falling out of the Disk can either send you into interstellar space or down, down, down the gravity well of our system.

At least the comets give us some warning when they go rogue: as they approach the inner system, the heat from the sun vaporizes their iced-up gases and molecules, leaving a tail stretching millions upon millions of miles, always pointing away from the sun.

On their newfound orbits these comets may not hit any inner system planets, but they will try, try again, looping back century after century. Once plucked out of the Disk, they tend to set up new regular routes that can send them looping around the sun for thousands of years. Some of them are even famous, like good old Halley (it took a long time for Earth astronomers to realize that comets can be regular and dependable visitors, and Halley was the first to crack it by predicting the return of his eponymous cometary friend).

Eventually, though, the comets break apart; they weren't exactly solid, after all, and weren't meant for life this close to their parent star. As they disintegrate orbit after orbit, they leave behind a cloud of meteoroids to scrape against the atmosphere of any intersecting planet. Not really a big deal, if you avoid them. Thankfully we've already charted the major leftover comet-clouds, so you should have plenty of warning.

But there are always new comets trying their luck with a loop around the sun and the chance to visit the fragile worlds of the inner system. By all accounts the Disk should've have exhausted its supply of comet wannabes long ago. There are only so many spare parts in the solar system's junkyard, and one by one they get nudged away to death by fire or a death by ice, never to bother another traveler again. Something must be refreshing this icy reservoir, and that can only come from something even further out.

✪

Past the Scattered Disk it gets even more mysterious, and even worse. You see, the Main and Kuiper Belts are only an issue if you're traveling in-system, along the plane. Since the Kuiper and Disk are past the orbit of any major planet, you don't have much reason to pass through them unless you're on some especially unlucky trajectory to your next destination. And the Main is only a problem if you're traveling between the inner and outer system. If you just want to escape altogether, your best bet is to loop your way up or down perpendicular to our home system altogether.

But no matter which direction you go, if you want to leave the system and pass into interstellar space you have to make it through the Oort Cloud. Like its name suggests, it's not a ring or a disk, and it's not just confined to the plane of the solar system. It surrounds in all directions.

For ages, we didn't even know the Cloud existed. In fact, to date no astronomer or explorer has ever directly observed or encountered a member of the order of Oort. There's plenty of raw material thought to be here—enough to make a copy of the Earth five times over, but all that stuff is spread out over such a tremendous volume that it's hard even to comprehend.

By the time you reach the inner edge of the Cloud, you're already a few *thousand* AU from the sun, and by the time you're through it, you're a quarter of the way to the nearest star—a solid year's worth of light.

So five Earths seems like a lot. But crush up those Earths into bits no wider than a city block and spread them in a giant shell a light-year thick. This far from the sun, even the shiniest objects are barely going to glitter, so the Oort is as mysterious as it is vast.

We only know of the existence of the Cloud because it's a source of particularly odd comets. While some comets come from the Disk and have regular, predictable orbits, returning to the scene of the crime every few decades or so, some comets come once and only once, appearing from random directions in the sky. Tracing their orbits backward reveals an origin thousands and thousands of AU away.

Random direction in the sky at great distance + a spherical shell delivering these comets to us = the Oort Cloud.[9]

Unlike the Main Belt, the Kuiper Belt, and the Scattered Disk, which are all home to at least one decent-sized roundish object, no known planet wannabe exists in these depths, only small icy death balls lurk here.

This far out, the sun's gravity is barely a suggestion; the Cloud stays intact simply because it largely has nothing better to do.

The Cloud is made of the final leftover bits of the formation of the system, either scattered out here during the early heady days, or left there because they never got a chance. They may even be the common bits left over from the formation of our local stellar group of stars and were never even associated with a single system.

Traveling through the Cloud on your way to someplace more interesting is, as with the Belts, just a matter of feeling lucky, punk. Since it's so vast, so thin, and so incredibly boring, you wouldn't even know you were in the Oort Cloud unless I told you.

The comets from the Oort are truly nasty. If you thought the Scattered Disk guys were cranky, you can imagine the bad moods these brutes have. They're barely even considered a part of the solar system, and they've been holding that grudge for millions of years.

Since the sun's gravity is so weak at these distances, any little kick, nudge, flick, twist, or bump can send comets diving deep into the inner solar system. A comet in the Cloud can spend billions of years hanging out, orbiting (in the very loosest sense of the word) some point of light that can barely be considered something more than just another star. It's cold and lonely, but at least it's home.

But all it takes is a few little nudges to change its mind. These nudges can come from anywhere; a passing star or a wandering molecular cloud can tug on their little icy heartstrings, shifting their orbits just enough to send them sunward. Even the galaxy itself can trigger the infall of a comet. Since the average number of stars on one side of our home solar system is different than another, there's an incredibly slight but detectable difference in the gravitational environments. Slowly, slowly, slowly, orbit after orbit, a random comet can get pulled by gravity further and further from the sun. But the way orbits work, the further a comet gets from the sun in the most extreme part of its orbit, the closer it gets on the nearside. In other words, a circular cometary orbit can get tweaked to become a long, skinny ellipse, and once that happens it's game over: down comes a comet into the inner gravity well to cause trouble.

Sometimes a comet from the Oort Cloud first gets deposited into the Scattered Disk, replenishing that unstable supply for a new round of mayhem.

The Oort Cloud surrounds the solar system on all sides, so these danger balls come from any direction, with any orbit. The distances are so great that in a human lifetime—and even a *humanity's* lifetime—the comets will make only a single pass before heading out into the galactic depths forever. Or sometimes they just smack into the sun on their first pass. Oops.

That makes these comets unpredictable, both in time and space. You never know when another one is on its way until it's already here and lights up a tail. By then, it may be too late.

Technically the Oort Cloud has only a finite supply of comets, and is steadily losing them one by one, either by expulsion from the system or through enough collisions to turn them to dust. But seeing as how the solar system has sported a Cloud for a few billion years already, I wouldn't hold my breath waiting for the fun to be over.

Comets: dirty snowballs, snowy dirtballs. No matter what you call them, they're bad news.

✪

There are a few other concentrations of asteroids and comets you should be aware of. Groups called the Trojans lead and follow Jupiter on its orbit, trapped in a peculiar balance of forces between that planet and the sun. The Earth has a convoy too, dubbed the Apollos. These groups won't do you any harm if you stay off their turf. If you decide to approach Jupiter or Earth, make sure you do it from a sensible direction.

And the rings. Right, the rings. Do I even need to say, "stay out of the rings of Saturn?" I suppose I do, just so I can sleep at night knowing I said it. Stay out of the rings of Saturn.

There, done.

So what are your best chances for survival? Despite their numbers, your chances of hitting any one asteroid or comet while moving in a small ship ought to be small: you have the sheer vastness of space on your side. The same vastness that makes it nearly impossible to travel from system to system, of course, but hey, you should count your blessings when you get them.

If you stay small, keep moving, and avoid the most obvious rocks, you should be fine. Should. Can't make any promises in this world. Over time you'll slowly collect scrapes and bruises from bits of dust and smaller meteoroids; that part's inevitable. You'll have to do regular maintenance checks and repairs if you want to do any long-haul voyages.

If you stay in one place too long, however, eventually time works against you and you're bound to get hit. So if you find a nice cozy planet to plant your flag on and call Awesome World, eventually a rogue asteroid or comet will come calling. You just have to hope that when it comes, it isn't a whopper.

I don't think you've developed a healthy enough fear of the dangers from these space rocks. Let's at least make one thing abundantly clear: you don't want to mess with them.

We can estimate the frequency and destructive potential of falling rocks based on craters left over throughout the solar system. Just look at Earth's Moon: big craters, little craters, craters over here, craters over there. Same story on Mercury and all the other airless worlds. The only craterless faces in the solar system are the ones able to resurface themselves through tectonic action or erosion (and I suppose the giant planets that simply eat comets for breakfast), but even they still bear some scars like pimples on a teenage face.

Since those naked worlds hardly ever change themselves and never put on makeup, their exposed surfaces are a visible record of the violence inherent in the solar system. It's from this documentation of damage that astronomers revealed the Late Heavy Bombardment and the subsequent four billion years of generally even strikes. Everywhere we look in the solar system, from scorching Mercury to frozen Pluto, we find battle scar after battle scar.

Our solar system was at its most frenzied in its youth, but even today it remains a dangerous place to live.

Want to see what kind of damage one of these comets can do? Earth astronomers got to witness an especially brutal bombardment by the comet Shoemaker-Levy 9 in 1994. That comet broke up into twenty-one pieces before slamming into Jupiter. Now Jupiter was big enough to shrug it off, but not without a few atmospheric disfigurements—the largest impact struck with a force equivalent to 600 times the entire world's supply of nuclear weapons.

We should be thankful that the King of Planets swallowed that broken comet before swinging into the inner solar system, where it could have put isolated, fragile Earth in its icy crosshairs. Over the eons Jupiter has served as an interplanetary goalkeeper, using its massive gravity to either deflect or absorb newborn comets inbound from the Disk or Cloud. In the past four billion years, who knows how many comets hit that giant world instead of us?

On the other hand, Jupiter's massive gravity also helps destabilize the Scattered Disk in the first place, making those harmless-in-place comets a danger to themselves and others.

So, Jupiter: it's a wash.

Here's what we know.

The Earth is pretty much getting rained on constantly, with dust drizzling in from deep space at a relentless but altogether harmless constancy. Just don't worry about the dust unless and until you start traveling through the depths.

The micrometeoroids are also always hitting us, especially during those events known to Earthers as meteor showers, when up to a hundred grainy and pebbly remains of a comet crash into our atmosphere every single hour. Again, only a danger if you don't have a thick protective atmosphere above your head.

The bigger rocks are obviously more deadly, but thankfully a lot less common.

A rock only the size of the room you're sitting in, traveling at the speeds they tend to do, has the energetic (and hence destructive) power of a typical nuclear bomb, and these guys hit a planet like the Earth about once every year. Why aren't Earthers noticing nuclear bombs going off all the time? Well, most of the Earth's surface is made of water, and nobody lives on the water, and depending on the angle of the strike most rocks just burn up and detonate in a flash in the upper reaches of the atmosphere. No harm, no foul.

But some hit land, and when they do, it most certainly is noticed. The largest recorded Earth-rock meetup happened in 1908 over Siberia (thankfully). There, a meteoroid a few hundred feet across managed to make it down to the last few miles of the Earth's atmosphere before finally giving up. It still managed to flatten trees across 2,000 square kilometers. That single strike, from a single meteoroid, could've easily bested a thousand simultaneous Hiroshima bombs.

And nature didn't even blink when she did it.

Bigger rocks—big enough to cause Serious Business (i.e., a civilization-destroying calamity)—are about a kilometer across and hit every 500,000 years, give or take. Hopefully take.

Extinction-level events are the ones to really worry about.[10] If your civilization is advanced enough, your species just might survive a one-kilometer rock. A good chunk of a nation will be literally wiped off the map with nothing but a smoking crater to mark the site, but there should be enough unharmed survivors on the rest of the planet to rebuild and move on.

But a five-kilometer rock? A ten-kilometer rock? Say goodbye to food and sunlight and hello to thousands of years of ash in the atmosphere. Fortunately these only hit the Earth every 20–50 million years. The last major one, 65 million years ago, wiped out a good chunk of all life on Earth with nothing more than an "Oops, sorry dinosaurs!"

When that thing struck, it was nastiness from day one. That meter most certainly made it through the atmosphere, hitting the shallow waters of the Gulf of Mexico just off the edge of the Yucatán Peninsula. The crater remains today, mostly buried under water. All 100 miles in diameter of it.

The meteor-now-meteorite punched itself a few miles into the Earth's crust, raising tidal waves a hundred feet high. Material from that impact—an ugly mix of Earth and meteorite—launched into space, reaching halfway to the Moon, before raining back down as molten birdshot, igniting forest fires around the globe.

The punch was powerful enough to literally shake the planet, triggering volcanoes to erupt on the opposite side of our home world.

Enough ash and dust kicked up into the atmosphere to blot out the sun for decades, plunging the planet into a never-ending frozen hell of winter. With little to no sunlight, the plants and algae—the bedrock of life's quest to turn sunlight into bizarre mating rituals—died off.

Don't be a dinosaur. Don't hang out on planets too long. When you're stuck to one place, you never know when your time is up.

✪

The above advice is especially true in younger systems. It takes some time to clean house after all the ruckus of forming a new system. Planetary collisions, migrating gas giants, gravitational breakups, the works. Not a pretty place, and certainly no place for an inexperienced explorer like yourself.

There's a wide variety of planetary systems out there, and each one brings its own unique flavor of danger. Some have gas giants barely brushing against the star, while others have three or more stars lighting up their host of worlds. While things like Oort Clouds, Scattered Disks, and Trojan systems are probably fairly common, you never know if a system will have an inner belt or a host of active comets. Always be careful, and stay alert, or get dead.

There are a few things you can do to minimize your risk. If you have to stay put on a planet, pick one with a nice thick atmosphere. All that air soaks up a lot of rocks. If you can, find a system with a large gas giant, as large as possible. It will act like a goalie, dragging rocks into its embrace with its massive gravity, before they pose a threat to more delicate worlds. Jupiter and Shoemaker-Levy, remember?

And please, pretty please, don't settle on an asteroid without first dragging it to a clearer orbit. You think that nobody would do otherwise, but there are a surprisingly large number of dim-witted explorers out there. Don't be one of them.

If you're out of choices, if a big rock is headed your way and you can't evacuate in time, there are still a few tricks you can play. Stopping a comet or asteroid is just out of the question. They're traveling at tens of thousands of miles per hour. You think you can counter all that kinetic energy with what? A puny nuclear bomb? Besides, the problem with most comets and asteroids is that they are not very well put together. Stick a giant bomb in their hearts and let it go, and what do you get? An irritated meteoroid, that's what. Most of the energy will simply escape through its porous body, and at best you'll just slightly reshuffle the insides.

And if you do manage to blow it up? Congratulations, now instead of one giant rock, you have thousands of medium rocks. A million tiny cuts can kill you just as well as a single blow.

You may not change a comet's mind, but you can, if you're clever enough, deceive it. If you know early enough—and this part is really, really important—it only takes a small course change to ensure that the rock goes sailing by at a healthy and respectable distance. A few megatons of explosives can do it, if you place them right.

But that only works if you have a good early-warning system. If the rock is too close, you just can't put together enough energy to give it a hard enough slap. Then your only choices are either to evacuate or hunker down.

Asteroids and comets are not completely known or mapped. They're not very bright. They move fast. Detection relies on scanning the sky night after night after lonely night, hoping to catch some glint of movement against the backdrop of fixed stars. While some asteroids have been flagged as potentially dangerous, because their orbits intersect the Earth's orbit (but not the Earth itself, otherwise it would just be lights out), as far as we can tell our home world is safe for at least the next thousand years or so.

But asteroid orbits are some of the hardest to predict. Things can change on a daily basis. Light pressure from the sun. Rotation. Extra special gravitational tugs from any of the planets. What was once a harmless asteroid can, without much warning, turn into a planet killer.

Alertness and constant vigilance—and some giant telescopes and sleepless astronomers—are a must.

There's a lot of *energy* out there in those rocks, and that energy can easily be converted into smashed ships or ruined planets. You can either be careful out there, or you can become just another pretty meteor shower. Your choice.

Solar Flares and Coronal Mass Ejections

That star looks friendly
Let's have a closer look
It wouldn't bite us, would it?
Not with that cute little—hey!
　　　　　—Rhyme of the Ancient Astronomer

B lack holes. Dying stars. It's a scary universe, and many travelers think that they can just hide away in a nice, cozy, mature system and stay out of harm's way. It's a shame for them that even normal-looking stars are a major pain in the neck. Your own backyard might as well be a pit of vipers for all the safety it gives you.

I'm talking, of course, about stars themselves. Even the sun. Yes, the sun. Nice, friendly, draw-it-with-a-smiley-face sun. It gives the planets light and warmth. Food for plants and on up the food chain to you. It's even in the name: without *sol* we wouldn't have a *sol*ar system. Sure, the system may harbor threats here and there: a rogue comet, a mean-spirited asteroid. But the sun? It's just . . . the sun.

Ages ago the sun had a violent youth and was best avoided. And in a few billion years the sun will expand to engulf the inner planets. Definitely don't want to be hanging around then. But its youth was a long time ago and its death is a long time in the future. The sun is middle-aged. Four-and-a-half billion years of regular, unbroken, boring hydrogen fusion. It has a steady job. A pension plan. A nice house in the 'burbs, a couple cars and a couple kids. It doesn't go to rock concerts anymore. No wild parties—it prefers Friday Pizza Night with a family-friendly movie.

Except when it gets . . . temperamental. It's usually pretty stable, but you should know that the sun occasionally has bouts of raging indigestion. The only way to relieve the pressure is to spew out a massive belch across the solar system; a blob of material from the sun's own furnace; a fireball of particles and electromagnetism.

These ejections scramble electronics and melt soft squishy things like your body. They disrupt communications and disable satellites. They're unpredictable and unavoidable.

Prominences. Solar flares. Coronal mass ejections. We have a host of names to describe the violence continually erupting from the surface of the sun. From the vantage point of a planet, tens of millions of miles away, a star like our sun is simple and bland.

But every star, including our sun, is a slumbering dragon, just waiting for the chance to awaken and begin breathing flame.

✪

Let's dig into the sun to see what's going on. Come on, it will be fun, I promise.

We'll start at the core, since it's at the center of everything. Here the crushing gravity of the sun's own weight is powerful enough to overwhelm the natural repulsion of atomic nuclei. And when it comes to hydrogen (which makes up most of the sun), a "nucleus" is really just a proton. A positively charged proton. And this presents a little problem. You see, protons don't exactly get along: put them next to each other and their electric charges will tell you "no thanks" and scoot away.

You could try to push two protons together yourself, but fat chance with your human strength of getting anything interesting done. But in the core of a star, the pressures are so intense and temperatures so extreme—we're talking 15 million kelvin here—that the protons begrudgingly get close together. And when they get close enough the strong nuclear force is able to take over, over- whelming the natural electric revulsion that the protons would normally face.

The protons are bound together now, whether they like it or not, forming a new element: helium.

Now here's a little weird random fact of the universe. Two protons glued together in a helium nucleus actually weigh less than two protons on their own added together. The deal is that it *takes* energy to rip the newly formed nucleus apart into two separate protons. Which means that two separate protons just hanging out have more total energy than when they're buddied up being helium with each other. And energy is mass ($E=mc^2$, remember kids?), so the helium has less mass than two protons.

So in the process of fusing hydrogen into helium, a little bit of energy escapes.[1]

Voilà, nuclear power.

That's it: that's the big engine that's been keeping the sun glowing for all these billions of years. Two by two the hydrogen atoms the sun was born with march into the core, get hitched, and make some helium. The leftover energy gets released as a flash of light deep in the core: a high-energy photon.

The actual chain reaction is a bit complex and depends on just how massive the star is, and in the case of our sun sometimes involves friends like carbon, nitrogen, and oxygen. It's also not straight hydrogen + hydrogen = helium, but involves some temporary combinations before the final product, and also spits out some weirdos like positrons and neutrinos. But you get the main idea: the sun crushes together hydrogen (much against its will, by the way) and makes a nice fat lump of helium. What it does with all that leftover helium is the story for another chapter. I don't want to overwhelm you with too many worries all at once.

Each reaction produces almost a vanishingly small amount of energy. But the core of the sun is large, has a lot of hydrogen, and repeats this process over and over again without breaking a sweat—around 600 million tons of hydrogen join the fusion party *every single second*.

In about three hours, the sun chews through the equivalent mass of the Earth's entire atmosphere.

Every second, about 4 million of those tons of hydrogen get converted into raw, pure, unadulterated energy. To put some human perspective on that, the most powerful nuclear weapon ever built by humans, the awesomely named Tsar Bomba, had a yield of about fifty megatons of TNT.

Line up two billion Tsar Bombas in a row and detonate them simultaneously. Congratulations, you've matched the sun's energy output for the time it takes you to blink.

Almost all life on dear old Earth looks to the sun as its ultimate energy source. Yes, there are some critters hanging out on the bottom of the ocean near deep-sea vents, but we don't talk to them much. If you're on the Earth right now, look around you. Look at the plants, trees, insects, mammals. Look at the food in your kitchen.

Most of the energy generated by the sun simply radiates away into the vastness of space, becoming just another sprinkling of starlight. The Earth is awash in only a tiny fraction of it. You, everything around you, and all the teeming life on our home planet relies on a mere 0.00000000217 percent of the total energy output from the sun.

But speaking of that energy and those nuclear reactions, it's kind of hard to vivisect (heliosect?) the sun to see what's going on in its guts. Thankfully, we have little messengers from the core that are able to whisper to us how all this nuclear fusion operates. It's these little guys called neutrinos, ghostly particles that are the silent byproduct of nuclear reactions, just like the ones happening in the solar core. The sun outputs a ridiculous number of neutrinos (billions upon billions are passing harmlessly through your body right now, and you don't even need to take a shower to wash them off), but we can detect them every once in a while with gigantic detectors on Earth

The neutrinos give us an intimate portrait of the solar core, something we would otherwise have a lot of difficulty directly measuring. (And who can blame our ancestors, before realizing that nuclear fusion was a thing, that the sun powered itself through combustion, or just retained heat from its formation. It's a hard problem.) But neutrinos are probably the Least Threatening Particle, Ever, so that's pretty much all you're going to hear about that from me (at least, for now).

You can imagine that it gets a little hot and sweaty down in the core. Fifteen million kelvin hot and sweaty, to be exact. And when you're at the center of the sun, you can't find a shady spot to save your life.

This hydrogen-smashing process occurs in a pretty big core, extending to roughly a quarter of the radius of the entire sun. It's a pretty abrupt transition the deeper you get into the furnace. Essentially, once you reach the right conditions, the fusion process turns on full throttle, no holds barred. Sure, there may be a sprinkling of random accidental fusion here and there throughout the sun, but the vast bulk of the sun's power is generated in a relatively small, tight, intensity of fire and chaos.

This means that the layer surrounding the core of the sun is dominated by the radiation produced from the fusion. Pack your sunglasses: the only way for all the energy to be released is to be carried away by the light. While temperatures decrease as you move yourself farther from the core, the differences aren't big enough to drive any interesting large-scale motions of plasma. Instead, it's just radiation piled upon radiation, transporting heat away from the nuclear core farther outward, as hard as it can.

That radiation has to put up a mighty struggle to move around, because even beyond the core the temperatures sit at a sweltering million or two kelvin, with densities of the hydrogen and helium soup to match.

Suffice it to say, there's a lot of bouncing.

You ever watch *The Price is Right*, and especially that one game called Plinko, where the plucky contestant has to drop a disk down an inclined wall, and watch in terror and/or delight as the disk bounces its way through the pegs, landing in essentially a random spot?

It would be so much faster—and less stressful for the contestant but less interesting for us—if the disk just slid down in a straight line. But the pegs block its motion, forcing it to follow an aggravating, ping-pong path.

OK, take that mental image and scale it down to the subatomic world, where the pegs are the jam-packed hydrogen and helium in the sun and the disk is the radiation trying to get through. And also scale this image *up* to a temperature of a million kelvin.

Good? Good.

It takes light basically forever to escape the sun, and this is the reason. A single bit of light—a photon—on average takes about a hundred thousand years to filter its way through these layers and out into the freedom of empty space. And just another eight minutes to make the hop to Earth.

Once you make it about three-quarters of the way to the surface, the physics changes in a big way. Think about it for a little bit (go ahead, I dare you): say you're a blob of hydrogen gas living your life in the outer layers of the sun. There's not enough pressure for you to start fusing, so you're happy to just hang around twiddling your thumbs. Beneath you is the nuclear furnace and the intensity of the radiation layer: a rowdy mess of fusion and photons, much too hot for your liking. Way above you, at the very outer surface, is the hard vacuum of outer space itself: much *much* too cold for your tastes.

So you've got hot on the inside and cold on the outside. What's your little chunk of gas going to do?

Well, what happens when you stick a pot of water on a hot stove? Stare at it long enough and it will start to boil, I promise. A random blob of water at the bottom of the pot, experiencing the closeness of the hot stove, will just happen to get slightly warmer than its neighbors. Hotter water means expanding water. Expanding water means less dense water. Less dense water means buoyant water. Buoyant water means rising water; like a hot-air balloon, that bit of water rises to the top.

I'm free, it says!

Then it reaches the surface and gets its dreams crushed: at the top, it touches the frigid cold air, releases its heat, immediately contracts, becomes denser, and sinks back down, its spirit withered. At the same time, its neighbors are experiencing opposite rides.

Back and forth, up and down, trains of water transporting heat from the bottom of the pot to the air above it. This is called convection, and it's one of the ways the heat can transfer from one place to another. It makes your water boil and it makes the sun boil too, but with great plumes of plasma instead of water for your pasta.

A boiling, roiling sun. Who knew? Besides astronomers. They've known for a long time.[2]

The temperature here on the surface is *merely* 6500 K, which is positively frigid compared to the core, but still, you know, hot.

Look closely at the surface of the sun. *Not with your eyes, sheesh.* You sure as spit don't see a plain featureless surface. Instead, you see . . . bubbles. Masses of overheated sun stuff a thousand kilometers across rising to the surface, touching space for the first time, thinking better of it and cooling off, sinking back to the depth where they belong.

Real scientists call them *solar granules*, but between you and me they're just bubbles.

This is the photosphere, the visible surface of the sun, where its energies are finally released, and light of all wavelengths escapes into the void.

That's right—all wavelengths. The sun emits lights in the visible spectrum pretty equally, with all colors of the rainbow well represented. But we think of our white sun as "yellow" because we have historically looked at it through the filter of a blue atmosphere, and white - blue = yellow. But once you get above the atmosphere you can see the sun for what it really is: a blazing incandescent ball of white-hot rage.

There's more. While the nuclear reactions in the core of the sun produce almost exclusively high-energy gamma rays, by the time that energy makes it to the surface, it's been bounced and scattered and absorbed and reemitted so many times that it's all mixed up, and all sorts of wavelengths come pouring out. In one of the most stunning examples of bad names, in the physics jargon this is known as *blackbody* radiation. The term has to do with the peculiar devices physicists in the 1800s used to study this kind of mixed-up radiation. And since they named it first, the term kind of stuck. Sigh.[3]

Anyway, the sun spits out a lot of different kinds of radiation. Infrared? That's the heat you feel on your skin on a warm summer day. Radio waves? You bet. Ultraviolet? Ouch, for sure, and while thankfully our atmosphere is able to block most of this harmful high-energy radiation, a few wavelengths are able to sneak their way down to the surface and make life tough for the skin of surface dwellers.

✪

But the solar fun doesn't stop at the surface.

Just past this chaotic cauldron is the corona. You normally don't get to see the corona, unless you cover up the bright surface of the sun with a special instrument or, for example, the moon during a total solar eclipse. I suppose the moon can be a pretty handy astronomical device. With the light from the surface blocked, you can see that the corona extends far from the surface. It's hot too; so hot that it doesn't emit much visible light, which is why it's so hard to see unless the photosphere is blocked. No, visible light is too wimpy for the corona. Instead, it's beaming out the good stuff: X-rays.

Despite being millions of degrees kelvin, you could swim through the corona without even noticing. (Note: you would still be burned to a crisp from the phenomenon known as *being so close to the surface of the sun*, but that's a different issue.) The corona is so tenuous and wispy that it barely registers. A single atom may be whizzing around from the extreme temperature, but if only a single hot atom hits you every once in a long while, you don't particularly notice. It takes a lot more than that to cook you, and the cotton candy tendrils of the corona just aren't going to cut it.

We're not exactly sure how the corona gets so hot. I mean, it's pretty strange when you think about it, which you should. The bottom layer of the corona is the solar surface, at 6500 kelvin. The outermost layer is . . . space itself, just a handful of degrees above absolute zero. And between these two cold extremes is the corona, at a million degrees. What gives? Just another deep and fascinating (and unsolved) mystery of the cosmos.[4]

As with most things in this guide, the corona is both beautiful and deadly.

It's from this region that the sun blurts a nasty poison into the rest of the solar system. How? Well, despite studying the sun since there have been people to study the sun, we don't really know. One idea involves magnetic reconnection.

Oh, did I mention the magnetic fields? No, I did not mention the magnetic fields. Right. Magnetic fields. The sun has 'em. Why wouldn't it? The sun is made of plasma, which is a fancy way of saying a hot soup of charged particles. The sun is also spinning. Charged particles moving in a circle make a magnetic field, but this magnetic field is not nice and neat and orderly *at all*.

The guilty party when it comes to this infernal tangling of the magnetic field isn't the corona itself—it's merely an innocent bystander to the nefarious machinations of the conductive layer beneath the surface. The crime story goes a little something like this:

The magnetic field of the sun usually looks like the magnetic field of any other charged-particles-moving-in-a-circle setup in the universe, including the Earth's: giant elegant looping lines poking out of the top end, flowing and wrapping serenely north to south, and poking in through the bottom end. Just like the field you might see by sticking a bar magnet in a bunch of iron filings.

(Side note for the truly nerdy travelers out there: since this kind of magnetic field has a nice simple structure with two poles, one north and one south, it's called a *dipole*. Monopoles are different creatures entirely, and we'll explore them later, because they're vicious.)

So the sun has this nice, beautiful, calm magnetic field, generated by its own rapid rotation. Sigh, if only it could just stay that way.

But alas, the sun doesn't rotate equally. It's not a ball of solid rock like the Earth, which forces all the parts of our home planet to rotate at the same rate. No, it's a plasma, and the middle bits of the sun rotate faster than the bits at the poles.

Take a bunch of parallel strands of spaghetti. Stick a fork in the middle and start twirling. You know that happens next.

The magnetic fields of the sun bunch up on themselves, getting stronger where they bunch and weaker where they . . . unbunch.

Then comes the convection. The constant up-down boiling of the convective layer drives its own mixing of the magnetic field. Like a second fork stuck perpendicularly in your pasta, it adds an extra complication to the tangling.

Over the course of years, lines of magnetic force twist and tangle and bundle and bunch in on themselves, driving them to incredible tension. Most of the time, most of the magnetic field lines stay safely inside the sun, but because of all the complex twisting, sometimes ropes and bundles will punch through the surface like a worm poking out of an apple.

If a rope of bunched magnetic field lines breaks the surface, it will stop the normal bubbling of gas in that puncture region. That's how we see sunspots: these are regions where the gas doesn't get to heat up like it usually does, because of the strong magnetic fields pushing down on it, preventing the normal flow of convection in its tracks. Hence, despite the fact that they're still several thousand kelvin, they look dark against their even hotter surroundings.

Then it's game time.

We'll get to the gruesome particulars in a little bit, but suffice it to say that for now there is a tremendous amount of built-up energy *released* (and that's the most welcoming euphemism I could think of) and after that release the magnetic fields settle back into their preferred parallel north-south arrangement, but this time with the poles reversed. But then the rotation and convection do their dirty deeds, and the cycle starts anew.[5]

We can track the behavior of the sun's magnetic field through the sunspots. When we see a lot of sunspots, we know that the magnetic field is reaching its breaking point. And conversely, little to no sunspot activity indicates a relatively calm, cool, and collected star.

This pattern repeats with surprisingly regularity. Earth astronomers have been noticing sunspots for thousands of years (recording them but not really understanding what they were seeing, which is a grand and ancient tradition in astronomy). But for the past few hundred years, we've been charting and mapping those blemishes, and they've found that it takes about eleven years for the sun to go from sunspot peak to sunspot peak.

Why eleven years, and not four or twenty-seven? We don't know, so we're just going to move on.

Strangely, the sun seems to be getting weaker. Over the past century, sunspot counts have regularly been dropping, with some years seeing no sunspots at all. Even the maximum activity of today isn't as severe as some of the minimums from yesteryear. What's going on? We don't know, so we're just going to move on.

However, even those powerful magnetic ropes don't have the energy by themselves to hurtle gas out into the solar system.

To get the right kind of magnetic field strength, you need to look right into the corona. Despite its tenuousness, it contains enormous amounts of magnetic energy, from all that twisting and convecting happening below it. And once those magnetic field lines tangle and cross, they build up an enormous amount of tension, like a pushed spring or an overstretched rubber band. And there the magnetic field lines will sit like a bad date going worse: straining, awkwardness, forced smiles, the tension becoming unbearable.

And then . . . the snap. The magnetic fields say, "I'm over it," and realign themselves to the way they prefer to be (that is, parallel), releasing the pent-up tension and energy in a blink of furious intensity. Millions of atomic bombs' worth of energy, in a single fit.

Anybody ever wind up a towel and give you a good crack? Imagine doing that with the force of a million atoms bombs at the surface of the sun.

You still with me? Incredible energies. Surface of the sun. Not happy fun times.

If you're lucky, the sun will just throw up a prominence: a gigantic bright arch of material, looping away from the surface of the sun and deep into the corona. Even the smallest could swallow an entire planet; the largest stretch across the very face of our parent star. These breakouts form in less than a day and can last up to a few weeks before settling back down into the surface. They are so large and intense that during total eclipses, they look like orange-red snakes writhing just underneath the hood of the corona.

But occasionally, when the sun has had an especially bad day and is particularly cranky, the prominences break loose from their confining magnetic fields. When they do, the loads of stored magnetic energy can go off in a flash: a solar flare of hard radiation, blasting the solar system and anybody unlucky enough to be in the line of sight with a lethal dose of X-rays.

But it gets worse. Sometimes when the prominences break the magnetic fields can slurp material up from the surface of the sun and launch it into space: a concentrated blast representing more energy than all of humanity's weapons combined. A coronal mass ejection (CME).

These aren't slowpokes either; blasting off from the sun's surface, they can reach the orbit of Earth in a matter of days. High-intensity radiation and a storm of energetic—even lethal—particles dredged up from the fiery depths of the sun. A deadly piece of star stuff flung in the direction of any unsuspecting planet, satellite, ship, or station.

A matter of days. That's all the warning you might have.

While there's definitely a relationship, the connection between sunspots, flares, prominences, and full ejections isn't fully understood, so it's difficult to build an early warning system. They're all certainly related to the magnetic activity of the sun, but whereas sunspots seem to be associated with internal magnetic fields breaking free, the flares and ejections appear to come from external fields plunging down into the sun like the hungry leeches that they are.

There seem to be more flare and ejection events when sunspots are at the peak of their eleven-year cycle. And ejections often accompany flares, but not always. Sometimes there are simple flashes with no ejections, and rarely there are ejections without any flares.

At its peak, the sun may launch a few ejections in the span of a day, while when it's in a more contemplative mood it will go up to a week between spasms. Sometimes the entire surface of the sun dims after a major event, as if the dragon needs time to recover before breathing flames again.

✪

Earth scientists assume that all this magnetic activity is to blame for the oddly intense heat of the corona, but heating up the corona to a hundred million kelvin—hotter than the core of the sun itself—a major solar flare releases *only* deadly amounts of radiation: lethal X-rays and gamma rays streaming through the system. You know how you have to wear a lead vest when getting X-rays at the hospital, and you're always a little suspicious that the technician runs away behind a big thick wall before turning the machine on? X-rays aren't meant to be taken in large does, because they can easily go from a "useful way to check for broken bones" to a "useful way to tumorize soft tissue." Not the easiest stuff to live with, but with proper shielding any fragile human cargo can be sufficiently protected.

Assuming, of course, that you have proper shielding. Layers of absorbing metals are heavy, and getting heavy things into space is hard. Well, not so much hard as expensive. And you think protecting your internal organs is worth the extra cost? Hope you like the taste of cancer, because every time there's a solar storm you're increasing your odds of catching one.

This radiation flies through space at the speed of light, since it's made of light, and that's its speed. By the time you notice that a flare or ejection event is occurring, you've already taken your poison pill. Flares happen more frequently than ejections, since sometimes the sun flares up without belching. Even if you're good at monitoring for coronal ejections, that won't save you from the flares.

Planetary denizens have it a bit easier: atmospheres are good for lots of things, including soaking up deadly radiation. A few miles of nitrogen can do wonders to the destructive potential of X-rays and gamma rays: they harmlessly knock around air molecules in the sky instead of harmfully knocking around your DNA. It only becomes a problem if you spend too much time outside of your security blanket.

This is less of a problem in the outer solar system: the radiation dose drops off with the square of the distance, so twice the distance means one quarter the intensity of nastiness. But the inner planets, especially Mercury, can be a wicked place. As for habitats or colonies on airless moons or asteroids, one would hope that habitat designers would place their structures deep underground, to avoid exactly this issue. Otherwise you would just be a walking talking cancer time bomb, waiting to go off.

But the coronal mass ejections cause trouble even for those smug planet-bound folks. Their atmospheres may protect them from the radiation, but the high-energy particles in the CME are another story. Those particles are a piece of the sun itself, snatched from the surface layers and tossed around like so many hardballs from the pitcher's mound.

From a drunken pitcher, that is. You never know where the next coronal ejection will be aimed. But given the fact they're usually larger than any planet, you don't really need to worry: eventually one will find you.

✪

We've got the radiation part of a flare and mass ejection covered, and unless you're adequately covered, the radiation's got you, too.

But then there are the high-energy particles, too. Protons, electrons, a few heavy nuclei raining down on your head with velocities a good fraction of the speed of light. They're called SEP. Solar energetic particles. *SEP*. Almost sounds cute, if it wasn't so deadly. The sun is constantly drizzling out a low-energy flow of these particles, but an ejection event carries a much more intense blast.

How do these little insane particles get accelerated? To explain it I'm going to throw a phrase at you. It's a whopper, so crouch low and lean forward a bit; you're going to get hit: *first-order Fermi acceleration*.[6]

Woah, nice one. Probably weren't expecting to find me saying that, were you?

Let's unpack that little gem. *Acceleration*: OK, it's a way of making things go fast. Check. (Enrico) *Fermi*: a dude who figured this out. Double check. *First order*: there are two kinds, and this is the simplest. Phew, that was easy.

Here's what's going on: remember that bit a little while ago about magnetic field lines snapping and cracking and releasing lots of energy? That energy comes in the form of a shock wave, like a sonic boom in the sun's atmosphere. That shock wave carries with it a changing magnetic field, or in other words, the magnetic fields on one side of the shock are different than on the other side. Normally nobody would care, except the sun's atmosphere is made of charged particles.

If a charged particle encounters a moving change in magnetic fields, it can bounce off of it, because of forces and stuff. After the bounce, it will be just a little bit faster. If there's only one shock, that's the end of the story, but in these complicated situations there are shock waves all over the place. Bounce, bounce, bounce, bounce; faster, faster, faster, faster.

And there you go: all these magnetized shock waves bounce around the sun's charged particles. Full of caffeine and all jittery, they fly off the handle and into space racing ahead in advance of a coronal mass ejection, capable of reaching the Earth in a matter of minutes.

So, to summarize. First you have the flares, or flashes of intense radiation. Bad. Then you get hit with the SEP, a burst of high-energy particles. Badder. Then comes the bulk of the coronal mass ejection itself. Baddest. A coronal mass ejection is a ball of charged particles weighing 1,000,000,000,000 pounds. Don't bother counting those zeros, there are twelve of them, all in a row. All those pounds accelerated to a tenth of a percent of the speed of light in a matter of hours from a magnetized game of ping-pong. Now I know that that speed doesn't sound like all that much, but considering the tremendous mass of one of these blobs, that's no small feat.

To the sun, that amount of material is belly button lint. To you, it's a heap of trouble.

Either way, as long as you're within an atmosphere, you're generally OK. Note that I said "generally" and not "100 percent." You're better off if your planet has a strong magnetic field, like the Earth does.

The same magnetic field that nudges around your compass can nudge around charged particles from the sun. Once they encounter a planet's magnetic field, they wind around them, streaming down the highways into the magnetic poles of the planet. There they charge (pun alert!) headlong into the atmosphere, where they tear electrons out of atoms and make auroras, the beautiful sky shows. Too bad you have to go somewhere cold to see them, unless there's a major event like a coronal ejection aimed square for the Earth.

When those happen, protected folks on planets get pretty light displays, sometimes even down in the tropics. Unprotected folks in space tell a different story.

✪

The last bit of fun with these death balls is probably the worst. I know, what could be worse than radiation sickness or cosmic cancer? Faulty electronics, that's what.

People on Earth first noticed this particular issue in the mid-1800s. Before then, there weren't any electronics laying around for people to notice when they went haywire. The ancient Chinese or Babylonian astronomers didn't have integrated circuits and telegraph wires, so they didn't have the chance to make the connection between "the sun is more active than usual" and "hey, my satellite isn't responding anymore."

In 1859, astronomers first noticed the sun's cauldron boiling over: the first recorded flare. At the same time, telegraph operators around the world were in for quite a shock: a literal electrical shock, zapping their fingertips. The aurorae were going crazy; strange lights in the sky where they shouldn't be. Magnetometers were reading off the charts. What was going on? Was the world ending?

No, the Earth had just experienced one of the largest possible coronal mass ejections, which turned out to also be the biggest ever known: the Carrington Event.[7]

The problem comes from those little charged particles. The atmosphere prevents them from slicing up people, but since they're charged and moving, they respond to magnetic fields. And since they're charged and moving, they *create* magnetic fields. Changing magnetic fields create electric fields.

A pulse of particles. A pulse of electricity and magnetism. An electromagnetic pulse. EMP.

Like a wave breaking on the beach, the EMP affects anything that relies on electricity or magnetism. Like, I don't know, *electronics*.

Those long telegraph wires were the perfect conduit: the EMP from that event shoved electrons around like a champ, busting circuits and shocking operators.

Nowadays we have a lot more of these *electronic* gizmos. You might be reading this on a device powered by electricity. Electrons whizzing around in their little conduits, doing math inside computer chips and lighting up displays. If a huge solar EMP hits, kiss that little device in your hands goodbye. Those electrons won't just whizz, they'll jump. They'll happily overload the carrying capacity of any wire or cell. Pop, sizzle, bang.

And the same goes for any other bit of technology.

Not everything, though: if the electronics are housed in a metal case, the EMP will just shove electrons around in the metal without affecting anything inside. If the electronics are built tough and can withstand temporary overloading, they'll be fine, too.

But, fortunately, we haven't yet had a chance to test any of those measures. Someday, though, we will.

It's worse in space. As usual, everything's always worse in space. Without that protective atmosphere, electronics feel the full brunt of the EMP attack. Memory chips get scrambled, solar panels get fried.

We've lost probes to such attacks. Poor little metallic souls; they never even saw it coming. The same sun that gave them light and warmth burned them up in a fit of rage.

Space agencies are constantly monitoring the situation, with probes located near the sun.[8] As soon as they see the familiar signs: a flare, a prominence, an ejection of material, they shout out a warning to the rest of the solar system. Satellites and ships hunker down in hibernation mode, their delicate electronics safer as long as they're unpowered. After the storm passes they reawaken, checking for internal bruises. They hopefully return to normal operations, dreading the next event.

★

So far I've used the sun as an example. It is the most well-studied and well-monitored star, and we are pretty dang good at detecting its mood swings and feeling out its temperament. We do this through a fleet of spacecraft and ground-based observatories constantly monitoring the sun for signs of solar weather.

The worst storms tend to come during the peaks of the eleven-year sunspot cycles, but don't let a solar minimum fool you, as flare and mass ejection events can come just as surprisingly during the downswings, too. Signs of a growing storm start with sunspot clusters—this is a clear indication that the magnetic fields in that region are reaching a dangerous breaking point. Prominences arching above the surface of the sun are a surefire giveaway that trouble is brewing. If we're lucky, the giant arch of plasma will simply simmer back down onto the fiery surface from where it came. If we're unlucky, it will snap, releasing a solar flare and maybe a coronal mass ejection not far behind it.

When an event is detected, a system-wide alert goes off. Any astronauts or space travelers currently out for a nice little space walk are ordered to come back inside to the safety of their spaceships. Satellites and probes must go into a low power mode, lest the EMP fry their little electronic brains.

A typical CME has a speed of a few hundred kilometers per second but are massive enough that they can easily engulf an inner rocky planet. Thankfully, we can see them coming from the flares that launch them, giving everyone in the solar system a few hours to a few days to prepare.

But that's the solar system, with its well-tested and long-running solar weather warning network. Other systems don't have that infrastructure, and you could be in for a wild ride.

Other stars are just as nasty as the sun. Some stars may be more active, especially if they're in their youth. Or if they're in their dotage. Or if they're middle-aged. Or if they're a different size. Or if they have a different composition.

Surprisingly, red dwarf stars can be some of the nastiest creatures out there. You wouldn't necessarily think it, though. After all, they're small, no more than half the mass of the sun. Their small size means that they don't pack a lot of gravitational weight, which means that the fusion reactions operate in their cores at a lower temperature. And lower core temperature means lower surface temperatures, hence all the redness.

And with lower energies and lower temperatures all around, they can't be *that* bad, can they?

Oh, they can. And they are.

The problem with the red dwarves of our universe (which, I should point out, are the most common kind of star you're likely to encounter in your travels) is that there isn't all that much distance between the core and the surface. With a bigger star like our sun, the naked energies of the core are surrounded by a thick envelope of plasma dominated by the radiation spewing out from the central furnace, and then *that* layer is surrounded by a massive cauldron of convection cells, before that energy can finally escape into the freedom of space.

But a red dwarf star doesn't have a radiation layer absorbing everything coming out of the core. The convection circulation directly connects the core to the outermost layer, and that same convection layer is also doing the lion's share of the work of powering the star's magnetic field. And as we've seen, magnetic fields can be friendly (as in the case of the Earth's) and nasty (as in the case of everybody else's).

Core closer to the surface. Strong magnetic fields.

This is a recipe for instability.

A red dwarf star can, within a matter of hours, suddenly surge 30 percent in brightness. And a few days later, it can hide half its face in giant starspots. And then it can flare out again—in one wild case, astronomers recorded a single flare ten thousand times brighter than anything ever seen from the sun.

Even our nearest neighbor, Proxima Centauri, is guilty of the crime of sudden and unexpected (and frankly rude) flare-up events. You just can't go *anywhere* in the universe these days.[9]

Compared to red dwarf stars, our sun is relatively mild, and our home world is pretty well insulated—and also distant. The Earth sits 93 million miles from the sun, which means any plasma-based temper tantrums have to travel a serious distance before affecting us. But habitable planets have to orbit nice and close to red dwarf stars—you can't get all that far away from them before it gets too cold for water to stay wet. This means that if you want to take a vacation to Proxima b, the not-fun-at-all name we give to the exoplanet orbiting our nearest neighbor—you'd best pack some sunscreen.

And at the other end, the most massive stars in the universe? Well, they're big. As far as we can tell, they don't flare up as often or as fiercely as red dwarf stars, but still . . . they're big. They're mean. They're powerful. A comparitively mild solar weather event from a giant star could easily overwhelm even the hardiest of defenses.

I guess you're just going to have to wing it. Without decades of detailed observations, it's difficult to know how a particular star might behave on any given day. Every star appears to have its own peculiar native star spot cycle, and every star has its own tendency to stay chill or to get irrational. It seems that younger stars have the most volatility—our sun certainly had a pretty violent youth, but beyond that, trends and general guides are hard to come by. Just stay alert for the warning signs, like increased star spot activity. Set up a monitoring station close to the star that can warn of oncoming ejections, if you can.

When in doubt, find yourself a nice thick atmosphere or a few layers of rock and hang on. If you're mobile, with enough warning it should be easy to avoid the major blasts of particles. But not the lethal radiation. By the time you see those, you're already nuked.

This universe throws a lot of nastiness at you, and sometimes you just have to stand and soak it up.

Cosmic Rays

Nip and tuck
Little knives do their work
Wearing down my bones
I should've brought a jacket.
    —Rhyme of the Ancient Astronomer

It's the little things that'll get ya. Oh yeah, sure, the big things will get you too—Mother Nature has boundless imagination when it comes to devising death traps. But there's a lot of little things out there too, and they can tear you apart, wear you down, and dissolve your organs just as easily as the big ones.

Take, for instance, cosmic rays. You'll encounter these guys and their friends an uncomfortably frequent amount of times in your adventures, so I'm sure I'll mention them again—they're kind of important—but it's time we dug in and got our hands dirty. I really need you to understand just how little these things care for you and your precious internal organs.

They're called "cosmic," so they must be in space. They're called "rays" because the Earth scientists who originally named them thought they were a kind of light, like gamma rays or X-rays (unrelated to stingrays and manta rays, by the way).

They were wrong, though. But who isn't wrong every once in a while? You should know the feeling: for example, you're wrong when you think you'll make it out in space.

Unfortunately for us, by the time everyone figured out just how wrong they were, the name had stuck. So now we've got cosmic rays that aren't even made of rays.

You'll get over it.[1]

They *are* a kind of radiation, though: the kind of radiation that still gives you cancer. We'll get to that in appropriately gory detail in a little bit.

Instead, these nasties are made of the same sorts of junk as everything else in the universe: mostly hydrogen, some helium, a dash of lithium, a pinch of heavier elements. Add a half a cup of electrons. Stir, cover, and place in million-degree oven. Ten thousand years before finished, remove cover and sprinkle with antiprotons and positrons. Remove when crust is golden brown, and particles are relativistic. Serve hot.

The "relativistic" in the recipe up there means they're moving at close to the speed of light. They're light enough that they can actually do it without too much sweat, unlike us humans, who heave and sweat just to get up to a decent jog. This gives cosmic rays quite a wallop: the most energetic ones pack a punch of around 50 joules. No, not a Joules Verne. *Joules* the unit of energy. One joule every second is a watt. You ever hear of those old-school incandescent lightbulbs? Every second one of those things delivers around 50 joules of energy.

So the highest energy cosmic rays have enough energy to power an old-fashioned lightbulb for one second.

That may not sound like much, but amazingly, it's definitely more powerful than any particle collider experiment that humans have ever made, that's for sure. Fifty joules is about . . . the total energy of a good baseball pitcher's fastball.

Oh, *still* not impressed? To understand just how much of a punch these things can deliver, remember that their energy is concentrated in a really tiny package. That energy is a combination of its speed and its size. You probably wouldn't enjoy it if you got smacked in the face with a fastball. If you made the ball smaller, to keep the same total energy, you'd have to make it faster. So, like, a bullet. I'm almost absolutely sure that you wouldn't enjoy getting smacked in the face with a bullet.

Now make it smaller, and faster. Even smaller, even faster. As small as a fundamental particle, and nearly as fast as the speed of light. Your face may not notice only a single one zipping through you, but your cells, your DNA, would.

That's a lot of energy in a tiny, compact package. It hurts.

Of course, most cosmic rays are not 50-joule monsters. Only a tiny percentage of them make it to that excited status. Most are much, much weaker, but even the weaker ones are more than powerful enough to slice, dice, and julienne fry. I know, I know, it still doesn't sound so impressive. But even the weaker cosmic rays have more energy than the most powerful particle colliders on Earth. You know, the ones that physcists use to probe the innermost workings of reality itself? Nature looks down at our most supreme efforts with the same parental sensation you have when your kid shows you their latest drawing of a dinosaur that looks more like a tangle of sticks.

How do they get so fast? So glad you asked, because I was going to talk about that anyway. You remember that buzzkill of a phrase "first-order Fermi acceleration?" Neither do I, so here's a reminder: charged particles can bounce from shock wave to shock wave, getting a speed boost every time. It's how things like the sun can blast off their mass ejections.

That was *first order* so now here comes *second order*. Pretty much the same idea, except instead of shock waves you have a group of clouds of gas (A flock of gas? A herd of gas? Suggestions welcome.) moving in the same direction. Say, from a supernova explosion pushing against a nearby nebula. If those clouds are magnetized, then any cosmic rays will bounce against those too, but with less back-and-forth as was the case for first order, and more of a zig-zag pinball effect. It's not as efficient as the first-order way, which is why it gets classified as second order.[2]

But the result is the same: the magnetic fields in the moving clouds of gas can accelerate particles to the required velocity. What's the required velocity? Fast enough to zoom all the way across the galaxy and into your spaceship.

Of course, with most awesome things in astrophysics, we're not exactly sure if this process actually works. I mean, it *works*, physically, but we don't know if it's responsible for most of the cosmic ray acceleration out there. As you might imagine, when it comes to some of the most powerful processes in the known universe—processes that can outclass anything we can set up in a laboratory—it's a little hard to figure out. But no matter what, cosmic rays are here and they're here to stay.

Some of the jargon words and phrases that I've tossed in your direction I encourage you to try out on dates or at parties. I think you should keep second-order Fermi acceleration in your back pocket though, and don't let it out. You wouldn't want people to think you're an astrophysicist, would you?

✪

But these aren't just atoms (protons, neutrons, and electrons in one happy family) of hydrogen and helium and all the others. They come from broken homes: at these energies, the atoms have been destroyed, the electrons have all gone, storming off one night in a fit of jealousy and anger, the nuclei left to fend for themselves. These cosmic ray particles are *charged*, and they're charging right for your lymph nodes.

And before you ask: no, we're not exactly sure where they all come from. I know, I know, in our journeys we will explore all sorts of high-energy events, like supernovae, coronal mass ejections, and quasars, and those are all potential sources of cosmic rays. But there are other possible sources out there too, and we're not sure who in this universe is in charge of making the most of them.

That's right: *in charge*, another space pun.

If you spot all the space puns in this book, you win a prize. The prize: you don't get to go into space and you end up living a long, healthy, productive life.

Speaking of the universe, these cosmic rays appear to appear everywhere, even in those incredibly vast distances between the galaxies. We know this because we see them in every direction we look. If they were only made in our galaxy, then we'd see them only when we look within the disk of the Milky Way. But we don't, so they don't.

Instead, every galaxy is a cosmic ray factory. See that small fuzzy distant galaxy in the sky? Riding along just behind all that pretty light is a crew of cosmic rays. All the high-powered events going on inside that galaxy are enough to pump out cosmic ray after deadly cosmic ray.

These events—whether the death of a massive star or the chaotic collapse of matter into a black hole—are powerful enough to launch particles halfway through the universe. Every firecracker in the cosmos that goes pop, sizzle, and bang releases deadly amounts of radiation, both the light kind and the particle kind. If you get too unlucky to be close to one, you're a goner, but even from far away—and keep in mind the mind-bendingly huge distances between us and the rest of stuff in the universe—their long deadly arms can still embrace you.

What's worse is that cosmic rays do suffer some losses in their journeys: they do run into stuff every once in a while, like all the clouds of gas floating around between galaxies, and sometimes they bump into a stray photon or bit of light roaming around the cosmos. It's estimated that once a cosmic ray enters the disk of our own galaxy—whether it was made here or came from somewhere else—it only lasts about three million years before sputtering out.

Why does that make things worse? Because it means that cosmic rays are *replenished*.[3] The universe didn't just make all the cosmic rays once in a spasm of insanity, otherwise all the cosmic rays would've disappeared eons ago. No, the insanity keeps on truckin'. The universe is a lot older than three million years, so if cosmic rays were only a Limited Collectors' Edition, we should've run out of them by now. But here they are, swarming like angry wasps, and that means that the cosmic ray party never stops.

Here's another perspective on what we're dealing with when it comes to cosmic rays. If you add up all the sources of energy in a galaxy, you get a few key players: the galaxy's magnetic field, radiation from starlight, and cosmic rays. Turns out they're all scoring about the same on the energy meter. If you want to impress your next date with a word they've probably never heard before, you can say that all these sources are in *equipartition*, meaning that they all contribute equally to the energy budget of a typical galaxy.

Equipartition. Cool word. My spell checker doesn't even recognize it, *that's how cool it is*. Go ahead, try it on for size. You feel smarter when you say it, don't you? Unfortunately, that means for every bit of energy from sunlight you soak up—on average, once you're well enough away from the sun—you soak up an equal amount of energy from cosmic rays. Think about that the next time to lay out on the beach to feel the heat. Yikes.

★

Speaking of which, the most prominent source of cosmic rays within the solar system is none other than that giant glowing ball of warm fuzzies that we call the sun.

Of course, in traditional astronomical fashion, since these kinds of cosmic rays come from a different source, they get a different name. We've already met the solar energetic particles, or SEP, the vanguards of a violent sun-based outbreak, which are yet another form of cosmic rays, but the sun also makes a different kind of cosmic ray, because why stop at one? In this case we call them the *solar wind*, which sounds light and airy and refreshing and not at all menacing. But the charged particles emitted from the sun and accelerated to high velocities are related to their cousins traveling in from more distant sources. A charged particle traveling at high velocities is a charged particle traveling at high velocities the universe around.

The sun is not a giant explosion from a dying star, so the particles our star is able to spit out are at much lower energies than those coming in from across the universe, but it's still bad enough to worry about.

And as with most things astronomical we're not exactly sure what generates the solar wind. I mean seriously, we barely have any clue what's going on here.

We do know a few things, but it's kind of hard to study because nobody likes getting that up close and personal to the surface of the sun and its scorching 5500 kelvin surface temperature. It's just not a fun place to do science, and so mysteries abound.

We do know that the particles that make up the solar wind originate not on the surface of the sun itself, but just past in the region known as the corona. As we saw earlier, the corona is basically the sun's atmosphere and has a temperature of over a million kelvin. That's hot, for those of you keeping score at home.

It's also huge. During a total solar eclipse, when the disk of the moon covers up the face of the sun, you can for a few brief minutes get a nice view of the corona. And if you do, you'll surely notice how it visibly extends to several times the radius of the sun.

Since the corona is so dang hot, the particles that it's made of are also hot (that's the definition of "hot"), and this temperature has a lot to do with turning some of those particles into the solar wind.

Fundamentally and microscopically, the hotter a gas is, the faster the little particles in that gas are wiggling around. It works for the air you're breathing and for the corona equally well. But not every particle moves at the exact same speed. For a given temperature, some particles will be slower than average, and some will be faster. And a select few overachievers will be *much* faster than average.

In the case of the corona, they can be so fast that they breach the escape velocity of the sun, and just decide one day to pack up their bags and leave home: the solar wind.[4]

This line of thinking is able to explain a good chunk of the solar wind, but astronomers are having difficulty explaining a) the sheer volume, b) the fastest speed, and c) how some of the solar wind appears to be coming almost from the surface itself, not way far out in the fringes of the corona where the gravity is weaker.

But since the solar wind is made of charged particles, magnetic fields are probably somehow involved.

Ah, magnetic fields. If an astrophysicist runs into a mystery, you can always bring up magnetic fields to solve the case. The go-to favorite suspect.

The strong magnetic fields in and around the sun can latch onto these particles in the corona and spit them out into deep space. Of course, the process is a little bit more complex than that, but that's the basic idea. Similar to how flares and mass ejections are launched, but on a particle-by-particle basis. And it doesn't just do this every once in a while, no, the solar wind is like a constant nonstop rainy day; a stubborn drizzle of high-energy particles that just won't give up.

For billions of years.

This solar wind completely saturates the solar system, sending streams of charged particles traveling at hundreds of kilometers per second, extending out far beyond the reaches of the planets (and even Pluto if you don't consider Pluto a planet).

However, eventually the solar wind starts to mix and mingle with the interstellar medium—the random bits of atoms and molecules and dust that just float aimlessly between the stars. The place where the solar wind meets the interstellar medium is called the *heliopause,* and most people who consider themselves experts on travel among the stars figure that the heliopause is the true boundary of our solar system.

Yes of course, at the distance of the heliopause (a little over a hundred times farther from the sun than the Earth sits), you're still feeling the influence of the sun's gravity, but if you're on an escape trajectory you could care less about how strong the gravity from the sun is at any given point—you're not attached, either physically or metaphorically, to the solar system. You care about what your local environment feels like. In essence, you're wondering how's the air out there?

But once you cross the heliopause, the universe surrounding you tastes just a little bit different. It doesn't *feel* like home anymore.[5]

The Earth is—for the most part—protected from the constant onslaught of the solar wind, thanks to that planet's magnificent giant force field—its magnetic field. Now even without the magnetic field the particles would still be mostly absorbed and blocked by the Earth's nice thick atmosphere, but it doesn't hurt to have that magnetic shield up at all times.

Some particles simply bounce off of our magnetic field, careening back out into the universe at some random trajectories. Good riddance.

Others are far too energetic and end up making their way into the atmosphere to wreak havoc. Good for them, I guess.

The slowest and least energetic solar wind particles—which I should mention are just the usual mix of electrons, protons, and sometimes something a little bit heavier—have the most interesting fate. When they begin to reach our magnetic field they respond to those forces, and charged particles really love to draw corkscrew paths around magnetic field lines. And so, they do. Depending on their charge, they travel either north or south along our magnetic fields, following them wherever they go.

And our magnetic field wraps around the Earth, puncturing our atmosphere near our geographic poles. And so that's exactly where these charged particles go, screeching into our atmosphere at incredible velocities. And as they travel, they strip electrons off of the molecules in the upper reaches of our atmosphere. Eventually those electrons make their way back to their atomic homes, and in order to do so they release a little bit of energy in the form of light. The end result: as all these charged particles get funneled into our atmospheres, they put on a little light show for us.

We call them the aurora. And they're very pretty.

The next time you see an aurora on our home planet, it's just Earth telling you that all defense systems are fully operational.

But some particles get trapped in our magnetic fields, just spinning around and around the Earth helplessly, forming belts of charges known as the Van Allen . . . belts. A typical spacecraft is more than well enough protected to sail through the Van Allen belts without any difficulty whatsoever. These aren't the worst characters in the universe, but let's just say you don't exactly want to spend a lot of time there, either. Below them, you're within their protective shell. Above them, you just soak in the normal background. Inside them? Anyone foolish enough to linger there is just begging to get an extra special dose of hard radiation.

Speaking of which, why don't we just go ahead and pretend that the entire Jovian system doesn't exist? I know you're tempted to go to Jupiter to see the Great Red Spot up close and personal, visit one of the mysterious icy moons, or maybe just watch some storm clouds roll past.

Don't.

Jupiter has the strongest magnetic field in the solar system, stronger than even the sun, creating aurorae that are a literally visible from across the solar system. The environment around Jupiter is so thick with these high-energy charged particles that even our most robust spacecraft can't stand more than a few orbits before succumbing to the damage. It's just a nasty place altogether.[6]

✪

So much for the solar wind, which, while related to proper cosmic rays, is quite a bit wimpier and relatively easy for planets like the Earth to shrug off. But what about the extrasolar baddies blasting in from the great cosmic depths? Is any force brave enough to stand up to the onslaught of these terrible particle hordes? Will any courageous power stand between us and certain doom?

Sure, why not?

Cosmic rays are charged, positive or negative, depending on the mix, but charged, nonetheless. That means they respond to magnetic fields: that's how a planet's aurorae work and how all that *Fermi acceleration* business gets started. Magnetic field lines act like highways for charged particles: once a little cosmic ray nears a field line, it can't resist taking the on-ramp and following the road. Well, most of the time; the route depends on the energy of the cosmic ray (if it's too fast it will just barrel on through) and the strength and arrangement of the magnetic field.

Our own Milky Way has a magnetic field; it's weak, tangled, but there. And that's enough to stop some percentage of the low-energy cosmic rays entering in from deep intergalactic space. The gas and dust floating around the galactic disk is good for something, too: it absorbs cosmic rays. It's not that great at the job, unfortunately, since it's all thin and spread out like too little butter on too much toast, but hey, it's better than nothing.[7]

It gets better the closer you get to a star. And worse. Better and worse. Better because a star's own magnetic field and its own stream of charged particles—the solar wind—form a bubble-like cocoon at the distance of about a light-year: the heliopause. Any cosmic rays passing in from interstellar space must first get through this barrier, and most don't. Up to 90 percent stop at heliopause, the outer gates of a star system.

Oh, I almost forgot the worse. It's because the solar wind—the thing making the heliopause possible and blocking a healthy fraction of those incoming cosmic baddies—isn't all that much fun, either.

Overall, it's best to stay within the protective cocoon offered by the sun's wind—or the wind of any star, for that matter. In deep interstellar space, and especially intergalactic space, it's just nonstop cosmic ray after cosmic ray. While it's not entirely pleasant to be soaking up the solar wind, it beats the alternative. Barely.

Any other sources of magnetic fields help, too. The Earth has a strong one—surprisingly strong, if I do say so myself—and it too can trap and guide cosmic rays, just like it does the solar wind.

But if you can't find a good solid reliable magnetic field once inside a star system, if you want to be protected from cosmic rays, you need either miles of rock or even more miles of gas. Whichever security blanket you pick, make sure it's a nice and thick one.

The Earth's atmosphere does a pretty good job of protecting that planet's fragile life. Mars, not so much. Venus, wow, good luck getting *anything* through that soup. So yeah, if you really *really* hate cosmic rays, but love lead-melting temperatures and pressures, Venus is the place for you.

The Earth's atmosphere notices these cosmic rays; sometimes they run into ozone and tear it apart. Ozone is really great at soaking up harmful UV radiation, so it's a little bit melodramatic: as the ozone protects surface dwellers from cosmic rays by absorbing the high-energy charged particles, it must sacrifice itself, making those same surface dwellers more vulnerable to high-energy light.

It's also possible that cosmic rays are ultimately responsible for lightning, providing the initial spark when they knock a few electrons loose in their groundward journey through a thunderstorm.[8]

After all is said and done, after the galactic magnetic field, the heliopause, the Earth's radiation belts, and the atmosphere, cosmic rays still get through to the surface of our home planet. For every square-meter patch (say, the easy chair you're sitting in), you get a dose of about *ten thousand* of the lowest-energy ones every single second of every single day. And night. And time of year. As far as we can tell, the rain of cosmic rays is pretty much constant. An unending light drizzle instead of brief thunderstorms.

A poisonous drizzle though. Acid rain.

When it comes to cosmic rays, the preferred scale for measuring their strength is something called the *electron volt*. It's an incredibly technical and obscure unit, used only by the most physical of physicists, but it's worth sharing so that you can at least pretend to know what you're talking about. One electron volt (shorted to eV) is how much energy a single electron gains when being accelerated by a single volt of electricity.

The term was invented for use in high-energy particle colliders, so you can see why it applies to cosmic rays, which are the result of nature's own atom smashing experiments.

Anyway, the lowest-energy cosmic rays, the ones that are currently slamming into you a few thousand times per second, have an energy of about a billion electron volts.

Cosmic rays a thousand times more energetic hit you about once per second.

Cosmic rays a thousand times more energetic than *that* come at you about once per year.

A single cosmic ray a thousand times more energetic than *that* will hit you in your lifetime, assuming you stay safely under the blanket of the Earth's atmosphere for your entire life, though at those energies, the atmosphere really isn't all that useful anyway.

The total amount of cosmic ray radiation accounts for around 13 percent of all the radiation a typical earthbound human will experience in their lives.[9] Other sources of radiation include radon from the ground, medical imaging, and bananas. I'm not joking about the bananas, but this book isn't *How to Die in Your Kitchen,* so I'll leave it at that.

So there you go: even if you live your whole life inside the protective layers of an earthlike planet, you still get an extra dose of radiation from the cosmic rays, equivalent to an extra X-ray exam every few years. Usually when we do X-ray exams, we balance the risk of slightly increasing your cancer risk with, I don't know, figuring out what the heck is wrong with you right this very instant. But this is an X-ray exam that you didn't ask for and you don't need. It adds up, year after year: there's an extra chance that a random cancer on Earth was caused by a star dying on the other side of the universe.

✪

Astronomers love a good mystery, and nature sure does know how to deliver them. And when it comes to cosmic rays, there are a couple puzzles for us to ponder.

I'm sure you can go along with the idea that, generally speaking, there are tons of low-energy cosmic rays, relatively fewer medium-energy ones, and then the extremely energetic ones are understandably less populous in the universe. Physicists use something called a power law to describe this kind of distribution, and generally there's a very predictable relationship between the energy of a cosmic ray and just how popular it is.

Except when there isn't.

There are a couple little hitches in the relationship, where things aren't exactly lining up the way they should. In general, to keep it as broad as possible here, there are slightly more medium-energy cosmic rays than we might naively expect (and we really are naive here, so give us some slack) and again unexpectedly more of the ultrahigh-energy bullets.

Of course, where some astronomers see a mystery, other astronomers see an opportunity. They think that these deviations in the relationship between cosmic ray energy and frequency (which are called, for various obscure and slightly silly reasons, the "knee" and the "ankle") give us a clue as to the ultimate origins for all cosmic rays.

We think that the lower-energy cosmic rays come from within our own solar system, probably launched off the backside of coronal mass ejections and solar flares (yes, this is in addition to the normal solar wind drizzle). The medium-energy cosmic rays seem to come from our own galaxy. Supernovae go off every few years somewhere around here, and we can possibly measure the rate of cosmic rays hitting us to figure out just how common supernovae are. Neat.[10]

But the extremely high-energy cosmic rays? Oof, now those are something else. We humans think we're all that and a bag of chips. We have giant and impressive particle colliders and we're smashing atoms all over the place, revealing the innermost workings of the subatomic world. We reach these incredibly high energies and convince that ourselves that we've got it all figured out.

Cosmic rays impact the Earth's atmosphere with an energy about a trillion times higher than our most powerful particle accelerators. On the regular, Mother Nature shows us who's really boss.

As a slightly ranty but necessary aside, this is why nobody should be worried about microscopic black holes. Every once in a while, you'll hear people mutter under their breath with some vague concerns about particle accelerators and experiments creating some tiny black hole that will end up consuming our home world and presumably them along with it. First, microscopic black holes are nothing to worry about, because they tend to evaporate instantly (assuming they even exist in the first place), and if they do manage to survive longer than a picosecond, calculations suggest that they would take a few billion years to grow up to be the size of anything worth worrying about.

Second, if microscopic black holes are produced in high-energy particle collisions, then our atmosphere has been serving as a black hole factory for the past . . . well, as long as we've had an atmosphere, which is about four billion years now and counting. If nature is capable of making a microscopic black hole, then guess what folks, there's already one there at the center of the Earth.

Nature just doesn't mess around.

Anyway, the most powerful cosmic rays are almost comically energetic, reaching energies over 10^{20} electron volts. To put that in perspective, a proton with that energy is traveling over 99.99999999999999999999 percent the speed of light. Of course, physicists have managed to come up with a clunky name for them, calling them ultrahigh-energy cosmic rays, abbreviated UHECRs (and I'll just go ahead and call these "you-heckers" because that's about what it feels like). The first you-hecker was found in an experiment in 1991,[11] and it was so powerful it was called the OMG particle, as in Oh My God particle. (Not to be confused with the poorly named "God particle," the Higgs boson. It's just a really powerful cosmic ray, OK?).

You-heckers pack a real wallop, so much so that we actually have a bit of trouble understanding exactly where they come from. Part of the problem is that they're incredibly rare; even our most sensitive and largest detectors only see them every few years or so. It's a problem without a lot of statistics, so it's hard to find a culprit for them out there in the universe.

For the few that we found, they're not tied to any particular galaxy or place in the sky. They just show up being jerks. We do know that they come from outside the galaxy at least, because they appear in all directions in the sky. If they were somehow launched within our Milky Way, then they would appear to be . . . coming from the Milky Way.

They don't, but past that we're a little bit lost. One of the biggest challenges is that cosmic rays coming in from the greater universe first have to tangle with our weak-but-still-present galactic magnetic field. But what our galactic magnetic field lacks in strength, it makes up for in volume. It's huge—it's literally the size of our galaxy. So cosmic rays, even super-duper energetic you-heckers, will feel the subtle deflections of that magnetic field, altering their trajectories once they come in from the intergalactic void, and by the time they hit the Earth their true origins will be completely obscured.

But in all honesty, you-heckers shouldn't reach us. At all.

For these most powerful cosmic rays, the options are really limited as to their sources. Supernovae are impressive, and we'll get into gory detail later, but even those aren't strong enough to generate these kinds of energies.

Really the only things in the universe that reliably pack enough energetic punch to accelerate tiny particles this close to the speed of light are the regions around giant black holes known as active galactic nuclei.

Now active galactic nuclei (aka quasars and blazars, and we will definitely get to all this later) are certainly places to avoid in general. But thankfully, as we will explore when we get there, for various cosmic reasons they're all incredibly far away from the Earth.

Too far.

You see, you-heckers have a hard time making their way through the universe. The problem is the cosmic microwave background, this cold radiation leftover from the early, more interesting days of the history of the universe. These photons just simply soak the universe with temperatures barely scratching above absolute zero. But when you're traveling at nearly the speed of light, those photons take on a slightly more energetic character.

It's like a pool. If you're just swimming slowly and efficiently, you can get across the pool pretty easily. But if you go too fast, that water essentially acts like concrete, and if you've ever experienced the bright-red burn of a belly flop, then you know exactly what I mean.

The upshot is that the very existence of the cosmic microwave background sets a limit to how far you-heckers can travel before their energy peters out. After about 150 million light-years, they stopped being you-heckers and turn into regular cosmic rays. But almost all the active galactic nuclei are much further away than this. So it appears that the one source in the universe that's powerful enough to generate you-heckers is too far away to generate you-heckers.

What the heck?

It's honestly an open mystery. Perhaps there's some sort of exotic supernova that can, every once in a while, generate the right amount of energies. Maybe there are nearby galaxies that sometimes become temporarily active, flaring far too quickly for human observations to have caught them in the act. Maybe it's something even more mysterious.

Wherever you-heckers come from, you best stay the heck away.

✪

Let's talk about showers. Since we're familiar with each other's habits by now, you're probably pretty sure that I'm not talking about a spraying-water-to-clean-your-stink shower. Very good, you're learning. We're talking about showers of particles.

A planet's atmosphere may do a good job at stopping most cosmic rays, but the story doesn't end with "stopping a cosmic ray." These things are high-energy, and when little particles get to high enough energies weird things can happen. If given sufficient motivation, they can transform into other particles.

Yes, you heard me right: one second you may have a proton, and you blink and the proton's gone and in its place you have a stream of pions, muons, neutrinos, and gamma rays. No, I didn't just make up those names. Those are real particles. Maybe they're not garden-variety particles, but they exist and everything.

Most cosmic rays are protons, but a "proton" itself isn't just a single particle. It's made of other particles—*quarks*—glued together with, bear with me, *gluons*. It's a ball. A messy blob of stuff. Don't think of a proton as a single entity, but rather more like a biological cell. Things can be made of protons, just like you can be made of cells, but the story doesn't end there: the cell is made of other stuff, and the proton is made of other stuff.

"Is that where it ends? Are quarks and gluons made of other stuff, too? Just how far down does this rabbit hole go?" Don't get smart with me. It's not a danger to you, so I shouldn't bother explaining, but no, as far as we can tell, quarks are the bottom of the barrel.

Now back to work, kid, you've got more important things to worry about.

So a cosmic ray proton can go speeding through the cosmos until it hits, say, a molecule in an atmosphere. If it has enough energy (and guess what, they do), it's just like what happens in a particle collider: *boom*. The proton gets blown apart into its component quarks, and these quarks recombine in interesting and artistically expressive ways.

Certain rules do have to be followed; it's not all willy-nilly particle fever. The total charge before and after has to stay the same, for example, as well as a few other things. But even with those restrictions there can be quite the party. The proton can become a set of pions, which are very unstable, and they quickly break apart into streams of muons, neutrinos, and high-energy gamma rays. The gamma rays themselves can split into pairs of positrons and electrons, and all these high-energy products can continue to hit innocent molecules in the atmosphere, creating an ever-widening branch of new particles and radiation. A shower. Eventually either all these products fizzle out in the atmosphere, or they hit the ground.

Just like what happens in the remnants of a supernova, except in your friendly happy-go-lucky planetary atmosphere. Or the metal wall of your spaceship. Or your brain.

Here's a fun fact. One of those products in the shower is the muon. It only has a lifetime of a couple microseconds, which even near the speed of light isn't long enough to hit the ground from the upper atmosphere. But thanks to the time-stretching effects of relativity, its internal clock slows down, and it has all the time in the world to wreak its tiny little havoc.[12]

✪

Let's talk about clouds. This time, I mean a friendly kind of cloud. I promise.

Let's say you're traveling out in some unexplored region of space, and you're starting to feel a little nauseous. You're worried about cosmic ray exposure, but your cheap off-the-shelf detector quit on you a few light-years back, so what are you to do?

Easy. You will need:

- A small aquarium
- A source of bright light
- A thin metal plate
- Plenty of duct tape
- Clear silicone sealant
- 100 percent pure isopropyl alcohol
- dry ice
- felt liner

You did pack all that, right? Those are essential spacefaring supplies anyway, so you should have them stashed away somewhere.

Attach the felt to the bottom of the aquarium and soak it with the alcohol. Place the metal plate onto the dry ice, and flip the aquarium over and onto the metal plate. Shine the light into one side of the aquarium.

Wait about fifteen minutes.

The bottom of your newly constructed *cloud chamber* is very cold, because of the dry ice. Since the top is at room temperature (note: do this in a room), the alcohol will come out of the felt and settle as a misty cloud at the bottom.

Eventually, you'll see some tracks zoom through. Almost like a meteor shower, but made of cloud instead of fire. As a cosmic ray or one of its muon children pass through the cloud, they may hit one of the alcohol molecules, ripping its electrons away and ionizing the molecule. This now-charged molecule starts looking for some friends, and that's all that's needed for that bit of the cloud to condense from a vapor into a liquid. Since the charged molecule got a kick from the cosmic ray, it quickly zooms through the vapor.

If you put a strong magnet underneath, the paths will curve, just like they do from the Earth's magnetic field.

With enough observations, you should be able to measure just how much trouble you're in.

There are of course other ways to detect cosmic rays. Bring a strong enough magnetic field with you and something for the cosmic rays to slam into (besides your head), and you can trap as many as you want.

Also, the cosmic rays are so fast that when they pass through an atmosphere, they're going faster than the speed of light in air. Pay careful attention: they're not going beyond the maximum limit, which is the speed of light in a vacuum, but they are going faster than light in air, which is totally legit. When that happens, they emit a ghostly light called Cherenkov radiation.

The fastest way I can describe the physics behind Cherenkov radiation is that it's almost like a shock wave when something travels through the air faster than the speed of sound, but for light. If you have a lot of air handy, you can look for this telltale signature.

Or you can do what Earth astronomers do and build giant vats of pure water and look for the flash in that. Some astronomers even took it a step further and built a detector out of the biggest batch of pure water they could find: the Antarctic ice sheet. Right at the south pole, at this very minute, detectors buried within a cubic kilometer of ice are hunting for the frenetic passing of a cosmic ray.[13]

✪

Usually people freak out whenever they hear the word "radiation." But you're smarter than that: you know that there are different kinds of radiation. Some radiation is "ionizing"—it can rip electrons off of atoms and do some serious damage. But other radiation is "nonionizing"—it just generally waves at you. For example, microwave radiation is nonionizing. Microwaves can still be dangerous: I wouldn't stand too long inside a microwave oven because the water in your body will start to boil, but it's not going to rip apart your DNA.

It's the ionizing kind that's troublesome. Ionizing radiation can be high-energy light or high-energy particles. Guess what kind of radiation cosmic rays are? Ding, ding, ding, we have a winner: it's ionizing.

Usually molecules are happy just playing their little chemical dances, but when some ionizing radiation comes along, it rips the electrons off of one of the molecules, making the molecule positively charged, just like in the cloud chamber. A positively charged molecule is in berserker mode, a bull in a china shop, a toddler with too much sugar. You get the idea. They make and break new chemical bonds, they bind things together that shouldn't be bound together, and they tear things apart that are meant to stay put.

If it slices up your DNA, the next time your cell goes to copy itself, it gets an error. Then another error, and other. Sometimes these errors just peter out and the cell dies a miserable death. Sometimes these errors make the cells go haywire, and haywire cells are called cancerous cells.

Even without slicing up DNA, cosmic rays can still be dangerous. The oxygen you use to breathe is already pretty reactive and chemically interesting, which when controlled makes for some really handy biological processes. But some oxygen ripped up by cosmic rays becomes extra reactive, binding to anything it can find, like cell walls, and rupturing them. Get enough cells broken apart and there you go: tissue and organ damage.

Cosmic rays harm electronic processes, too. Everything going on down inside a computer is based on the flow of charged particles, aka electricity. Lots of charge: a binary 1. No charge: a binary 0. And all else follows. Add some cosmic rays into the mix and things that were supposed to be 1 are now 0, and things that were supposed to be 0 are now 1. Whoops.

A typical home computer experiences a few errors every month due to cosmic ray strikes. It's worse in space: the Voyager 2 probe once suffered a malfunction, and the most likely cause was a single bit getting flipped from a stray cosmic ray.

Of course, now we're smart and we put in error-checking routines for just this sort of nonsense. But these routines don't catch all the errors, just like your DNA isn't always able to smooth over troubles all the time. Computer cancer. Is that a thing? It just might be.[14]

✪

The worst part of cosmic rays—and let's admit it, there are lots of horrible parts already—is their insidiousness. Unless you receive a huge direct and lethal blast, you won't feel anything. Your fingers won't tingle, your skin won't sizzle, you won't see lights before your eyes. You just go on ho-hum about your life, sitting at home on a planet or colony or off exploring the universe.

But bit by bit, cosmic ray strike by cosmic ray strike, your DNA starts to fail. Most of the time it doesn't matter; the cell just dies. But the more invisible strikes you suffer, the greater your odds of hitting the jackpot. The cancer jackpot.

Am I using the word "cancer" too much? No, I'm not. This is serious and deadly stuff, and it's a reality of space travel.

A typical human on Earth experiences around 3 millisieverts of radiation per year. Milliwhat? Millidoesn't matter, it's a unit for as much radiation exposure you can get. If you want a baseline, there it is: 3 millisieverts is the normal everyday background of your life, and doesn't really cause any problems.

Airline pilots, crews, and passengers get about double the exposure of folks on the ground, especially if they fly over polar regions, where the Earth's magnetic fields channel the cosmic rays. While it usually doesn't cause health problems, it looks like crews have an increased risk of getting eye cataracts, another consequence of this horrid stuff.

So think about that next time you're on your way to a tropical getaway: little cosmic rays invisibly slicing through your body. Zing. Zing. Zing.

An unshielded human in space, away from the protective layers of atmosphere and magnetic fields, gets about 1000 millisieverts in a matter of days. With shielding, they can spend up to six months before the dose reaches 100.

Even shielded, a short interplanetary hop, 180 days from Earth to Mars, delivers 500 millisieverts. Another year on the Martian surface, without a magnetic field and only its wimpy thin atmosphere, gets you up over the 1000 millisievert threshold, which if you're counting is in the "serious trouble" range.

Cosmic rays come in different energies, and as we saw the slower ones are much more numerous; this is actually a *good* thing, because they're more likely to hit multiple parts of the same cell and just kill it, instead of just striking DNA to cause cancer. It's the high-speed ones that are precision cancer-causing machines.

It's estimated that Earth to Mars travelers, without proper shielding, would lose about 5 percent of their cells to slower cosmic rays during the voyage. Five percent, including skin cells, heart cells, and precious brain cells.

If the sun throws a coronal mass ejection event at you when you're among the planets, it ends quickly. Instead of just slowly raising your risk of generating a cancer, it gives you *acute* radiation poisoning, and within a few days or even hours your internal organs simply fall apart.

First comes the nausea and vomiting. Then the diarrhea. Maybe just a bad stomach flu, right? But then comes the severe headaches and the fever. Nasty, but survivable, right? Then the tremors, seizures, and lethargy, as your central nervous system shuts down. Then mortality.

Fortunately, that's the worst case. If you've set up your warning system properly, you can avoid the mass ejections or hunker down in heavily shielded environments.

Speaking of shielding, there's good news and there's bad news. The good news is that thin plates of metal are enough to stop most cosmic rays, along with gamma rays and X-rays, which also cause these kinds of issues. This is really handy, because since we only need thin sheets, we can get that protective armor into space easier.

The bad news is that thin sheets of metal are good at stopping rays. What happens when the atmosphere stops a cosmic ray? It produces a shower of radioactive and energetic particles. Great, instead of one bee you have a whole swarm, and if the shielding metal is too thin, that shower goes right through the hull and into your heart.

To stop this, you can try a few things. The simplest thing is to have thick enough walls to stop the cosmic ray *and* the shower particles, but that's expensive and heavy and impossible to get into space, so that's never gonna happen. You can use thick layers of something light, like some gas, to absorb the shower. That doesn't weigh a lot, but it's hard to get it under control. Finally, you can wrap your ship in strong magnetic fields, just like planets can do, to deflect the cosmic rays.

But those are all experimental ideas. Right now, the recommended strategy is, unfortunately, "shut up and take it."

PART TWO

INTERSTELLAR THREATS

Stellar Nurseries

New life, new brightness!
But youth and violence.
Like good wine and cheese,
the best stars are aged.
    —Rhyme of the Ancient Astronomer

B irths are messy events. Beautiful, sure. A wonderful part of the natural cycle of life, of course. A reassurance that a new generation will carry the torch, without a doubt. But also messy. Very, very messy.

So it goes for people and so it goes for stars. Some people want to be there at that special moment. To be in the room, to hear the screams, to feel the excitement, to cut the umbilical cord. Other people are content to wait outside and take a look at the newborn after it's had a good wipe down.

It's especially true since the births are shrouded in light-obscuring clouds of black dust. I'm talking about stars now, by the way, not people. It's actually very hard to witness the birth of a new star, and explorers and travelers are always eager to witness such momentous events.

I mean, a new star! Can you imagine? There may be 300 billion relatives already roaming around the Milky Way, but each one is precious. A beacon of heat and light, shining out against the eternal darkness that pervades the universe. A source of warmth. A potential home for planets. For life.

What kind of star will emerge from its cocoon? A tiny red dwarf, burning feebly but steadily for tens of billions of years? A massive monster set on a path straight for supernova? A sunlike star, with a wide habitable zone, ready to host some rocky planets with liquid oceans? A binary pair? Quintuplets?

It's a moment of excitement, with maybe a little trepidation. But no matter what, it's a moment to be witnessed, recorded, remembered, and celebrated.

It's also dangerous, so yeah, don't bother.

✪

Like most things, stars start in the places that are not stars. The vacuum. The void. The emptiness. The nothing. In order to bring the light, you must first start in darkness. It could almost be some prehistoric myth, but in our case it's just physics and chemistry doing their thing. And time. Lots and lots of time.

Most of our galaxy—and any other galaxy—is just a hot, thin soup of random atoms. Mostly hydrogen, because most of everything in this universe is hydrogen; a good fraction of helium, the perennial silver-medal winner in all matters astronomical; lithium, which we don't talk about much because honestly, who cares about lithium; and then all the rest. "All the rest" being the entirety of the periodic table of the elements.

Side note: astronomers are a funny bunch when it comes to nomenclature, as we've already seen. In the case of the stuff that will eventually stuff itself down to make stars, astronomers classify the whole universe into three and only three elements: hydrogen, helium, and "metals." To be fair, if you're not named hydrogen or helium, you make up no more than around one percent of all the raw material in the cosmos, so you would be considered nothing but an annoying rounding error if it weren't for all this interesting chemistry that you could participate in.

Anyway, the hydrogen and helium come from the earliest moments of the big bang itself, and as violent as those epochs were, they are safely in our past and thus do not present a danger to the modern-day traveler (except for possible exotic fossil remnants, like cosmic strings, which we'll get to later).[1]

Thus if you were to pick up a random atom from the deepness of interstellar space, there's a solid chance it will start with the letter H.

From there, dead and dying stars pollute/enrich (depending on how much you care about "metals" in your analysis) the so-called interstellar medium with all the heavy stuff, from carbon and oxygen to iron and potassium. We'll talk about how stars go about doing this in short order, but for now just know that stars are made of some mixture of primordial (though by no means pristine) elements left over from the big bang, and a bunch of guts spewed out from previous generations of their kin.

Gross, but it's the circle of stellar life.

Sometimes the little atoms, like the especially promiscuous carbon, get frisky and start sticking together. Ever so slowly (because space is mostly empty and mostly cold), microscopic dust grains will form. Now, they're not immediately harmful, but they play a role in our story, so I wanted to give them a proper introduction.

There's a lot of free-floating bits of junk out there; enough to manufacture around 5 billion new stars, if our lazy galaxy ever got around to it. Unfortunately for the Milky Way, our star-forming heyday is long over. Nowadays, we barely croak out a few stars every year. Basically retired but still showing up at the office out of habit, I guess.

While that's a lot of stuff to make stars, it's spread out over a pretty decent chunk of space: the entire 150,000 light-year width of the Milky Way. That makes this the thinnest soup you could possibly imagine: densities vary anywhere from the low end at a hundred thousandth of a billionth of a billionth of the air density at sea level to the high end at a staggering . . . ten thousandth of one billionth of the air density at sea level.

Even at its densest, the interstellar medium beats out the best laboratory vacuums ten thousand times over.[2]

So as you might imagine, star formation is a pretty rare event, which is a bummer if you're foolishly trying to witness it, but a relief if you're sensibly trying to avoid it.

In order to make a star, the gas and dust floating around all willy-nilly in the galaxy has to get really dense, really quick. It also has to chill out, a lot. Did I mention the temperature? A few thousand kelvin, typically. The interstellar gas is so thin you would still need a jacket if you swam through it, but it's still technically hot, and hot things don't like to clump together into small spaces.

In your interstellar travels you're mostly going to encounter a whole bunch of nothing (we've covered this), but every once in a random while, you'll find yourself sailing through an ever-so-slightly-denser-than-average clump of gas and dust: a nebula.

A nebula is a cloudy, wispy thing, and it takes its name from the Greek for "cloudy, wispy thing." Some nebulae (or nebulas, take your pick, I won't mind) come from stars turned inside out: planetary nebulae from the death of stars like our sun (we'll get there in a bit), ejection from supernovae (we'll get there in a bit), remnants of kilonovae collisions (we'll get there in a bit), and so on. These tattered remains eventually disperse and intermingle with the general galactic background, forming nebulae of various shapes and sizes. The death of a star, as brilliant and moving as it is, serves only in the long run to pollute our interstellar waterways with heavy elements.

For our starbirth-hunting purposes though, we're interested in the densest and the coldest of the bunch: the *giant molecular clouds*. There's a breakdown going down here.

"Giant": it's big. Not the biggest thing in the universe, but compared to a peanut or a planet, quite large. A few hundred light-years across, and enough stuff to build a million suns, give or take. That's pretty beefy.

"Molecular": it's made of molecules. That right there should give you some clue that something interesting is happening. Molecules in our universe are actually a pretty rare thing. You might think they're common, since you're made of molecules and you breathe molecules and you eat molecules and you swim in molecules. But you're wrong. Most of the stuff in the universe is a high-temperature plasma, with the electrons ripped from the atoms in a big hot soup. Where's most of the stuff in the solar system? The sun. What's the sun made of? Plasma. Why is it a plasma? Because it's hot.

Anyway, back to our definition. The giant molecular cloud is cool enough to let the electrons back into atoms, and the atoms to join up together into molecules. Mostly hydrogen, because most of everything is hydrogen, but also some carbon monoxide, ammonia, dust grains, and some other characters.

Finally, "Cloud": it's a big blob, and "Giant Molecular Big Blob" doesn't sound very technical, so "Cloud" it is.

Giant molecular cloud. "GMC" if you want to sound snazzy.

They're pretty dense places, as far as things go in a galaxy. Denser than the interstellar medium, and even the galaxy on average, but less dense than your head. And that's saying something.

These molecular clouds are pretty chill places, in both senses of the word. They're very cold, almost absolute zero, and they're not in too great of a hurry to do much of anything. Left to their own devices, they can stay pretty much as they are for a quite a while. But this being the galaxy and all, nothing is left to its own devices for very long.

It's here, in the giant molecular clouds, where we have our best chances to catch the formation of a star. There are other sites, other cold and dark patches of nebulosity, like the amusingly named Bok Globules.[3] Maybe not amusing enough to take the top spot, but definitely in the running. They are also called small molecular clouds, but that's not nearly as fun. They only have enough material to pop out a few dozen suns at most, and they're typically no wider than a single light-year (in galactic terms they might as well be nonexistent), so if you're eager to witness the momentous event of the birth of a star (against my advice), then your best bet is to simply pass by these and head for the giant clouds.

These kinds of dark nebulae are pretty easy to spot. Look around. See all those pretty, glittering stars? Notice any obvious black ugly patches? Chances are there's a dense, dark cloud in that direction, blocking the light from the stars behind it. If you can see these clouds, you can't actually see them. Just their silhouette. Steer into the blackness and eventually you'll find yourself in the thick of the action.

You can't go too far away from the midplane of our galactic disk—most molecular clouds are found in a band only 500 light-years thick. When our galaxy makes stars, it only does it at the razor's edge.

Besides looking for where there aren't stars, the nebulae are emitting light of their own, mostly from the emission of infrared light from hydrogen buddied up with itself. So slap on those high-powered night vision goggles and you should see nebulae aplenty. You can also tune in your radios: the dust scattered inside them emits weak radio waves. I don't know what music they're playing but it's definitely not static.

Altogether, though, the dark nebulae are pretty hard to spot, but we have built up a decent collection of known star-forming regions. Some of the nearest include the Rho Ophiuchi cloud complex and the Taurus Molecular Cloud, just over 400 light-years from the sol system. The Orion Nebula, visible to the naked eye on Earth as a "wait, that looks a little too fuzzy to be a star" and famous through the ages of human history, can manufacture over 2,000 suns and sits over 1,300 light-years from the sun.

Astronomers and navigators are constantly updating the galactic maps when new dark nebulae and star-forming complexes are identified. Please make sure you have the latest updates, especially when exploring the mid-plane of the galaxy, to make sure you can chart a safe course.

✪

I would say that the most dangerous phase of star formation is when the clouds first begin to collapse and fragment, but that would be a lie. It's the second most dangerous.

A random nebula can hang out being a nebula pretty much forever; it's in a state of graceful equilibrium between its own gravity wanting to pull it in on itself and its own heat and pressure wanting to diffuse it, known as *hydrostatic* equilibrium. "Hydro" because this vaguely has to do with any kind of fluid. "Static" because it's the same. And a state of equilibrium, how poetic. With all the forces in balance the nebula is simply a nebula; no more, no less. But then something happens.

What is the something?

It can be a few things. Sometimes the nebula can randomly acquire too much mass (say, by the interstellar medium casually raining onto it), making its own weight overwhelm the supports put in place by its own pressure. Instability ensues—named the Jeans Instability after the Earth scientist who first worked out the math. We'll see this same game played out again at the ends of the lives of stars, with gravity the ultimate victor, but here it's a creation story. A violent one, but a creation nonetheless.[4]

Or the collapse can be triggered by some outside agent. A shock wave from a nearby supernova. Passing close to another cloud. A clump of dark matter (dark what? We'll get to that later) swings through. There was the cloud, minding its own nebulous business, still in equilibrium, when some pest disturbs it.

Sometimes star formation in one cloud can trigger star formation in a neighbor through the ejection of massive outflows and shock waves. The neighbor then goes on to trigger star formation in its neighbor, and so on and so on in a nuclear-fueled game of telephone.

Slowly, ever so slowly at first, the cloud begins to contract. This process can take up to a hundred million years and this giant complex, once stretching hundreds of light-years across, begins to fold in on itself, the densities inching higher. This game lasts so long because it's really, really hard to compress a giant cloud of gas. To make itself small, the cloud has to dump a lot of heat into the vacuum (because cold gas is denser, yes it's as true for weather on the Earth as it is shapeless, tenuous structures in the galaxy), and the only way to do that is through the emission of radiation (just like you if we spaced you). But the contraction tends to heat things up through friction, which takes time to release out in radiation, which allows another bit of contraction, and so on and so on for ages upon sleepless ages.

And then the fragmentation begins.

The party starts and ends with turbulence, one of those wonderful natural phenomena this is totally ubiquitous and yet almost entirely unexplained. Like the nature of time. Time and turbulence—it's said that if you can understand both of those, then you've unlocked the ultimate secrets of the universe.

Good luck with that.

Right, turbulence. Suffice it to say you don't want to be in a giant molecular cloud when it starts getting a little rough. A fluid (and when it comes to turbulence anything from an ocean surf breaking on a beach to these great nebulae are considered "fluids") is naturally a little bit sticky—the atoms and molecules like to hang around together. We call this natural stickiness the "viscosity," and different fluids will have different viscosities, because reasons.

But a fluid also has kinetic energy—everything from microscopic wiggles to macroscopic thunderous flows. When the kinetic energy gets too high, it overwhelms the natural stickiness and the fluid, for lack of a better term, simply falls apart.

Turbulence is a) chaotic, meaning it's entirely impossible to predict exactly how it will shake out in any given circumstance, and b) hilariously violent, as I'm sure you've encountered when traveling through a turbulent atmosphere. When a fluid goes turbulent, the excess energy tends to flow from larger scales to smaller ones, until it reaches a point where the energy is low enough in a certain patch that the viscosity can finally take control again and maintain order.

When it comes to giant molecular clouds that are undergoing catastrophic collapse, the onset of turbulence leads to a fragmentation of the cloud—the once-uniform giant nebula now finds pieces of itself pinching off, disconnecting from everybody else, and marching to the beat of their own, independent drum. Like friends who have kids and then effectively disappear from your social circle, these bits might as well be in their own universe. Clumps, tendrils, sheets. The gas cloud quickly starts to show a very complicated and rich structure.

Within each one of these little nuggets a cocoon begins to form, thick and rich like a molecular milkshake. The embryos continue to cool as well, but once they reach about 170 K, the clouds can't compress any more. Emitting infrared radiation was the key to dumping out all the excess heat, but now the core is too thick to allow light of any form to escape—it's too pinched off. The core once again realizes that blissful state of hydrostatic equilibrium, with its pressure and temperature balancing the inward pull of gravity.

But layers outside those cores continue to fall in—they can release all the heat they want. As they crash onto the dense core, shock waves ripple throughout, increasing the temperature of the core as they go. Eventually the temperature reaches an incredibly high and dangerous threshold: 2,000 K. At that temperature, the once-molecular hydrogen breaks apart into free atoms, turning the gas in the core briefly transparent again, allowing radiation to escape and the core to return to the tricky cycle of cool-compress-heat-cool-compress-heat.

Smaller and smaller. Denser and denser. With heat able to escape, the core can compress, and with the higher densities, the very center of that core can reach previously unimaginable temperatures, yet again achieving hydrostatic equilibrium, but this time at a blistering 20,000 K.

A protostar is born.

✪

And that's pretty much the deal: the gas continues to collapse, forming a dense ball at the center. That clump starts to get hot enough to glow and gets a new name: a *protostar*.[5] The protostar isn't burning hydrogen in its core quite yet; it needs a few more million years to mature. But it becomes incredibly hot and bright. All the falling gas has to give off its energy somehow.

It's shortly after this young not-yet-a-real-star phase that the newborns are most violent. They are extremely active and turbulent, crying all the time and throwing tantrums, knowing what they want without knowing how to ask. You can sympathize with the young star. It's hot and dense, but not hot and dense enough to have well-regulated fusion. Clumps and streams of new gas are continually falling in. It's rotating rapidly. Some of the gas inside it is ionized and turned to plasma, which means charged particles are moving, which means magnetic fields are adding their sinister influence to the mix.

And it's hungry. The surrounding gas quickly collapses to form a thin disk, which funnels new material into the protostar. Suckling from its now-razor-thin nebula, the protostar continues to grow. But with that hunger comes violent crying outbursts. As the material swirls inward, it spins up strong electric and magnetic fields, which can twist and warp the gas into serpentine paths.

Some of that gas reaches the still-growing protostar. But some gets locked into lines of magnetism that instead wrap around the star and shoot outward from the poles, racing upward and downward at a healthy fraction of the speed of light.

One of the beautiful aspects of physics is that it's universal. The law is the law, from one side of the cosmos to other. If a physical situation has the right setup, you'll get the same result. Soon enough we'll meet the quasars, powered by supermassive black holes at the centers of galaxies, where incoming gas from an accretion disk launches into long thin jets. The physics that made that happen was gravity, magnetic fields, and a little bit of spin.

And here we have gravity, magnetic fields, and a little bit of spin. At a smaller scale than for the quasars (phew, otherwise our galaxy would simply be ripped apart, and which we'll talk about in gruesome detail once we get there), but the same deal.

The rejected streams of gas slam into the surrounding nebular environment, where it—for all intents and purposes—crashes into a dismal heap, giving rise to the curious structures with the even more curious names of Herbig-Haro objects.[6]

Take these Herbig-Haro objects as the warning signs that they are. Violence. Outbursts. Spasms. Energies capable of ripping holes in giant gas clouds. I know you want to witness the moment when the protostar emerges from its cocoon as a full-fledged fusion burner, but the chaos of these early epochs is not worth the trouble.

All that activity means that the comforting cloak around the new protostar isn't going to last long. Powerful winds from the young almost-star, jet outflows, and plain old-fashioned heat. It all means that the surrounding gas in the disk slowly boils away within a few short million years, either ionized and turned into a softly glowing plasma, or steamrolled out of the system entirely.

At last, the new protostar has made its debut.

When a cloud of gas compresses enough to trigger sustained nuclear fusion in its core, it's called a star. We're not there yet. And as stars evolve they follow a particular life cycle called the main sequence. We're not there yet, either. But now that the protostar has emerged and it has a new name: a pre-main sequence star. I know, I know, not the most romantic of names for this object that is about to become a beacon of heat and light for billions of years, but it is what it is.

If the PMS star—wait, wait, hold up.

Let me be clear.

PMS = Pre–Main Sequence.

Thank you.

If the PMS star is destined to be relatively low mass (say, no more than twice the mass of the sun), then in this stage of the pre–life cycle they are called *T Tauri*. That's not a stutter, just the name of the first of this kind to be found.[7]

You know about sunspots, but the star spots on these young objects can stretch almost halfway across their faces. Big spots mean big magnetic fields, which mean strong stellar winds. These winds launch hot bubbles the size of a solar system. There are brief intense bursts of radiation, too, mostly X-rays. It's everything you might normally encounter in a typical mature solar neighborhood, just cranked up to eleven.

Actually, cranked up to a thousand: the strongest outbursts are literally a thousand times worse than the flare-ups from the sun.

All this energy comes from the continued gravitational contraction of the PMS star. It's still not hot enough to achieve nuclear fusion, but the continued contraction and squeezing raises the temperatures higher and higher in the middle. As it tightens, the surface goes nuts, and now that the veil has been ripped away, it blazes clear and bright.

A young almost-star in this phase can lose almost half its mass in ferocious outbursts. Get too close, and you'll catch all that lovely star stuff right in the kisser.

More massive stars are just as violent and cranky, but are given different names, because as I'm sure you've noticed, astronomers just love to classify and sub-classify things. No matter what, these are some of the most extreme and tricky times in the evolution of the star. In some ways, their births are as messy as their deaths.

And then, after about 10 million years of violence and outflows and winds and turbulence and chaos, the innermost parts of the core reach a certain special critical threshold in temperature and density.

That density, by the way, is 10^{24} times the density of the interstellar medium from whence it came. And the temperature? Cracking an infernal 2.5 million kelvin. So dense and so hot that matter can't be itself under those conditions; the electric barriers between protons that keep them well away from each other are overwhelmed, forcing them together in the vice grip of nuclear fusion.

Hydrogen combined to form helium. A small release of energy. Repeated countless times over in the infernal core.

A star, born.

From the outside, nothing much changes. The surface is a little calmer as the source of energy—nuclear fusion—provides a stable, steady supply from the inside. The flow of gas from the surrounding disk reduces to a bare trickle, then to nothing.

Cleared of gas from the previous violent outbursts, the surrounding disk begins to coalesce and fragment on its own—we'll get to that part of the story in a little bit.

★

And this newborn star has siblings. We started with a massive cloud of gas a few hundred light-years across, with enough stuff to form a million suns or more. You think only one piece breaks off at a time to form a star? This identical process of breaking and collapsing is happening everywhere it can. If a shockwave passes through an entire giant cloud, then this can set off a dozen, a hundred, even a thousand starbirths.

A good fraction of the stars will be small—the size of the sun or even smaller. Others may form in binary or more-nary systems. And if a particular patch of collapsing gas cloud is big enough, and the conditions are right, a monster will be born.[8]

We're not sure how big the biggest star could be. We've definitely seen ones easily a hundred times more massive than the sun, but it's not clear if it was born with that mass or porked out on a steady diet of gas over a few million years. But no matter what, when the biggest stars are born, they're intensely bright, clearing away their protective cocoons and even driving away gas in the larger parent cloud.

By the time a smaller nearby protostar forms, its shell of gas is thick enough to prevent it from being eaten by the powerful radiation from its bigger neighbor. But any gas in between is sure to be etched away like water digging into a sandcastle. This is what gives star-forming clouds their intriguing shapes: a few bright, hot youngsters, already exposed, carving their names into the dust, with columns and pillars of gas shielded by dense clumps around protostars.

That's part of the allure: to sail among the pillars, to weave around the nebulous intricacies of the gas. But those big bright stars, the very ones that light up the clouds in beautiful displays, are constantly emitting X-rays, gamma rays, and cosmic rays. Much worse than your typical quiet, mature region of the galaxy.

Beauty can be deadly in these regions, especially since those massive stars are prone to go boom at any time. For the heaviest stars, the time it takes to form—a few million years—is about equal to their lifetime as a hydrogen-burning star. So by the time you enter into a molecular cloud, some of the bright stars lighting it up are ready to detonate. And as I'll emphasize as strongly as I can when we get to that particular hazard, you do *not* want to be near a supernova.

To make it worse, since the most massive stars will tend to form around the same time, they'll also tend to blow up around the same time. If one goes off another is likely to go off nearby. These supernova groups are powerful enough to launch bubbles of gas clear off the disk of the galaxy, driven by a burst of radiation and cosmic rays, like a cork popping off in a fizzy mess from a bottle of champagne. You definitely do not want to get caught in one of those—it's not as fun as champagne.

Eventually the entire molecular cloud evaporates or is blown away, and all the stars—even the smaller ones—have hatched from their eggs. You're left with a cluster, a family of a hundred or a thousand stars, ready to disperse into the galaxy and be a part of the community.

The first effect of all that motion is that newborn stars can be given large velocities in random directions. The cluster of stars that are born together will generally stay near their birthplaces for a few hundred million years before gently dispersing. But a few get shoved out earlier, either by chance interactions with another star, or blown away by a supernova. These "runaway" stars move through the galaxy up to five times faster than normal. They naturally wreak havoc: plowing through quiet neighborhoods, detonating in supernovae where they're least expected, the works.[9]

Giant molecular clouds are the unruly children in a galaxy. Eager, awkward, ill-tempered little brats. They may be fun to play with, or cute to look at, but they're nasty. They'll bite and scratch and hit at the slightest provocation. Stars are dangerous all through their lives, that's for sure, and they come from violent births and adolescence.

Keep your distance. Observe from afar. As the stars form, they light up their host clouds in beautiful ways, with colors and shapes unseen throughout the cosmos. But close up, they are hotboxes of radiation and turbulence, driven mad by the very stars they were trying to create.

★

The continued rounds of star formation are responsible for some of the most beautiful structures in the universe: the spiral arms of galaxies.

While the spiral arms appear alluring, glittering as they are and visible from across the cosmos, these are some of the most dangerous places in the galaxy (besides the core; whatever you do, don't go *there*, but we'll get to that). Despite their appearances the spiral arms aren't home to that many more stars than average. It's a little bit hard to measure, but it's only roughly about 10 percent denser than the gaps between. But they're so bright and luminous! What's going on?

Here's what's going on.

Galaxies have density waves. Just subtle ripples where the local environment is just a little more crowded than average, like the ripples of a wave on the surface of a pond. These ripples come from all sorts of things: orbiting globular clusters (which are almost utterly harmless and thus not a major concern for us), supernova outbursts, whatever.

These ripples aren't necessarily circular—they tend to be elliptical, because gravity actually likes to make ellipses more than it likes to make circles (for an example, look at the orbits of the planets in your very own solar system). These elliptical-shaped ripples are themselves orbiting around the center of the galaxy, because why wouldn't they do that. Since they're elliptical, they tend to pile up on themselves, and the pile-ups take the form of a spiral.

But still, these are only slight increases in density—a few more cars on the road, not a straight-up traffic jam.

But as these density waves ripple across the face of a rotating galaxy, giant molecular clouds pass into and out of them, which upsets and destabilizes them, triggering a new generation of stars. The first stars to peek through are the most massive ones, and those hot bright young stars are most easily seen from the other side of the universe. A coiled set of arms, a painting for each galaxy.

But those bright, hot stars live fast and die young, burning themselves out before you even get a chance to properly be introduced. That leaves behind the hangers-on: the mid- and small-sized stars that live on to a comfortable middle age and retirement. But by the time that they do—when their siblings have dispersed into the galaxy and started families of their own—they've moved past the density wave and are in amongst the gaps.

But since they're smaller and dimmer, they're harder to see, despite being almost as numerous as the population of the arms.

Imagine cars on a freeway approaching a traffic jam, and when they reach the jam they switch on their headlights. As soon as they fight and struggle their way through and are back on the open road, they switch off their lights.

From a distance, at night, you would only see the jam, not the rest of the busy road.

The spiral arms trigger star formation, but do not host a considerable excess of stars; just an extra population of the big and bright ones. But do not be fooled: the spiral arms are nothing to mess with. It's there where there are the most supernovae, the most intense bursts of radiation, the most fragmentation. All that chaos. The turbulence of the large molecular cloud, pinching and twisting and rolling. Blasts from the first generation of massive stars to go supernova. Outflows and bursts from young stars as they blow away their cocoons. It's an awful place, and that awfulness creates an echo that can be felt for billions of years.[10]

✪

Speaking of billions of years, our Milky Way galaxy can keep popping out new stars, hatching them mostly in the spiral arms, for billions upon billions of years at its nice and steady current pace.

But all that is going to come to a violent and quick end in a few billion years, because we're currently on a collision course dead set for our nearest intergalactic neighbor, Andromeda.

Yup, our two galaxies are going to merge. But it's not like an instant car crash, with a loud bang and deployed air bags and accusations of fault. There's a lot of space between the stars (because, space). Most of the stars will simply swing by each other completely harmlessly. But there are clouds of gas. And there is gravity.

The merging of the Milky Way and Andromeda will trigger a round of furious star formation not seen since the individual galaxies were first forming, billions of years ago. In a few short hundred million years or so, the newly merged megagalaxy will use up its entire supply of gas, popping off new stars like it's going out of style. Sure, it will *seem* like fun, but like an all-night bender, we're all going to wake up the next afternoon with a serious hangover.[11]

And no more gas to make new stars.

Once all the hot and bright stars burn off, the entire galaxy will grow dimmer and redder, never, ever making stars like it did in its glory days.

While the end state will simply be a sorry affair, at least it will be relatively safe. But when the galaxies are in the middle of their merger? Trust me, you want to find yourself on the opposite end of the universe if you can at all help it.

At least you've got a few billion years to prepare yourself.

✪

But the birth of the star isn't the end of our violent little story.

Oh no, it only gets . . . rockier . . . from here.

You'll appreciate the pun a lot more in about ten minutes.

Remember that cloud of gas and dust encircling our newborn protostar like swaddling clothes around a crying, burping, pooing infant? Soon after the protostar got its act together and compressed, the surrounding gas collapsed from a generic, vaguely lumpy cloud to a sharp, thin disk. Of course the physics of that process is frighteningly complex, what with all the turbulence, rotation, magnetic fields, stellar outbursts, and so on—a dangerous, dangerous place.

But you can at least attempt to understand this collapse, since nature likes to make thin disks as often as she can (seriously, the universe has basically only three shapes: lumpy, round, and flat). It goes like this:

- The cloud of gas is rotating around the protostar. Rotating things have centripetal acceleration (or centrifugal force, your choice), which makes the gas want to fly outward, but only outward left or right, not outward up or down. Just like when you ride a carousel, you don't feel any particular inclination to go fly up into the sky, just out to the sides. If we were looking at this system edge-on (and I'm sure you eventually will), the centrifugal force (or centripetal acceleration, your choice) wants to make the gas go sideways.

- The protostar has gravity, a lot of it, because it's the densest part of the whole thing. That gravity pulls *inward* in all directions, both sideways and up and down.
- The gravity that wants to pull inward from the side is balanced by the centrifugal force wanting to push outward to the sides. Net result: no movement in that direction.
- The gravity that wants to pull inward up and down has nothing to counteract it, so the cloud collapses in that direction.
- End result: the disk wants to get all snuggly with the protostar, but its own rotation prevents it from getting as close as it would like. It does indeed collapse, but into a disk.

And as that disk compresses, it spins faster and faster. Go find the nearest rolling office chair. Spin around with your arms out, then pull them in. You'll spin faster. The physics-types call that conservation of angular momentum. I call it fun.[12]

As the protostar is preparing to make its first debut onto the galactic stage as a T Tauri star (if it has the right mass range), the disk around it spins up, heats up, and gets very, very interesting.

It also gets a new name: a *protoplanetary disk*. The sudden appearance of the prefix "proto" in there should tell you something interesting is about to happen. Something that might, perhaps, lead to the formation of planets. Just taking a guess here. But how does a system go from "protoplanetary" to just "regular planetary?"

It probably starts with dust. The same static electricity of dust that makes it so hard to clean off makes it remarkably sticky, especially as interstellar denizens go. Tiny little harmless grains find each other in the swirling maelstrom, holding fast together amongst the incomparable forces of their environment. And more dust joins, and more, and more, and more.

Little tiny bits of dust, ignored throughout their entire lives as so much cosmic fluff, begin the journey of growing into planets.

The original molecular cloud that gave birth to this new system was cold enough to hold a lot of ices: water ice, ammonia ice, carbon dioxide ice. You name it, it's got an ice for it. All those ices were along for the ride through the collapse of the fragment and the spin-up into a thin, flat disk, but the vicinity of the not-yet-a-star is much too hot for them to survive. Very quickly, probably during one of the periodic violent outbursts, the ices nearest the young star evaporate.

But in the outer system they thrive and join along with their dusty brethren in building larger blocks. There's a line in every system, depending on the size and ferocity of the parent star, beyond which the ices can survive. It's called the frost line, and you should know that in our home solar system, it sits at about the distance of the asteroid belt. Past that line, the nuggets that will eventually turn into planets are much larger, much more massive than their star-loving fellows living further down the gravity well.

The first large objects are called planetesimals. They are only a few miles across and they are almost uncountable in number.

And just like the dust and ices that glued together to form them, they find and refind each other countless times over the course of millions of years. Building large, crashing and shattering, spinning, coming closer and receding further from their star. A chaotic, messy, careening, violent place, but one infused with the excitement and energy that can only accompany the formation of something new and different.

Slowly, steadily, the planetesimals increase in size. Inward toward the star, they can only get so big; there's only so much material in the universe that can withstand the heat that close to the star, and the inner portions of the disk quickly empty themselves out onto the larger—but not quite planet-sized—rocks orbiting the awakening star.

In the outer solar system things get big, fast. This is especially true early on, where there are tons and tons of raw hydrogen and helium gas still swirling around. The compound ice-rock cores are pretty gravitationally hungry, and in a miniature version of the growth of the central protostar they feed and feed on the surrounding gas, wrapping their dense cores in layer after layer of thick gaseous envelopes.

They become the giants.

But the violence of the growing star puts a cap on their maximum size. With every sudden outburst, more and more of the light hydrogen and helium is shooed away from the system, blown back out into the surrounding cloud or even the interstellar medium itself.

The most stunted of the giants end up further out, with hardly any of that rich gas supply to call their own, left to only feed on what ices remain, while the nearer giants gorge themselves on a buffet of hydrogen and helium. Hence the crafty astronomer identifies two types of colossi: the gas and the ice giants, like some fantastic creatures from a long-forgotten myth.

As the giants grow to hulking proportions, half in shadow in the outer reaches of the system, the inner rocky embryos collide and collide again, steadily growing ever larger in the process, one quaking interaction at a time.

By the time the star is ready to begin burning hydrogen in its core, the planets have already appeared. The innermost worlds are relatively small and rocky, sometimes managing to pull together a thin, tenuous you-seriously-expect-me-to-call-that-an-atmosphere. Beyond the frost line you have the land of the giants.

And past that are the ultimate leftovers: the tiny icy bits that never got to join in the planet party. The rejects. The sad-sacks. The hopeless. It's there they begin their devious plotting, before the star even begins to shine, waiting eon after eon for their chance to come hurtling back into the inner system, wreaking havoc on any surface-dwelling life that happened to gain a foothold. These comets were born with all the others, from the same primordial soup, but only know their star as an especially sharp and cold point of light set against the millions of others in the vastness.

★

It takes time for the planets to settle down and stop hitting each other, and the evidence of our history of violence is everywhere.

Something catastrophic may have happened to Venus to push its rotation down to a pace so low that you could easily walk on the surface and keep the sun in perpetual high noon.

Something catastrophic definitely happened to Uranus to knock it—along with its retinue of rings and moons—so far over that it sits almost perfectly perpendicular to the rest of the solar system.[13]

And Earth. Poor little Earth. The only inner world with a moon worth mentioning, likely formed from ripped-out guts after a Mars-sized protoplanet slammed into the still-molten Earth. An event so significant that we gave that rude intruder a name: Theia.

Now *that* event was something to see. Imagine that massive protoplanet ripping right into the young Earth, literally ripping out its heart. The combined planet instantly vaporized, spending the next few hundred years as nothing more than a ring of hot, ionized metals and rocks before regaining a sense of self and settling down back into a planet, but now with a newfound companion, the moon. And that moon now orbited a much larger, denser planet than what had originally formed at that orbit.

Even right on the cusp of the ignition of the parent star, when by all rights things ought to have really calmed down, the protoplanetary disk is still a rough place, with comets raining in from beyond the frost line, combo-punching the inner worlds. (This ultimately turned out to be a Good Thing, as those comets resupplied precious reserves of water that had been lost in the molten past. Next time you take a drink, raise a glass in toast to this so-called late heavy bombardment.)

And the migration. Wow, you thought birds could migrate, but have you ever seen a *planet* migrate?

Simulations (which are about all we have to go on, since we don't have time machines) suggest that the early days of our own solar system hosted many more planets than we have today, and two of the giant worlds—Neptune and Uranus—actually formed closer in before migrating to the suburbs, and there may have even been another massive planet that got ejected from the neighborhood altogether, doomed to wander the galactic depths cold and alone as a rogue planet.[14]

Other solar systems fare far worse. Sometimes their giant planets can get caught up in the whirls and swirls of the not-cleared gas around the star. What ensues is a complex dance as the disk itself steals momentum from the planet, forcing it inward as it continues to feed. Sometimes the giant planet falls into the star itself—a comparatively gruesome but nonetheless fast end.

Sometimes these huge worlds can get trapped close to their stars, even closer than Mercury orbits our home sun. They won't live long in that infernal embrace, the heat boiling away their atmospheres layer by layer. These so-called "hot Jupiters" were a surprise to early planet hunters.

Nothing in our own solar system even hinted that this could be possibility. But once we found world after strangled world, we had to completely rethink our understanding of planet formation. Again.

Planets migrate, bombard, suffer trauma after trauma. Like I said, births are messy things. That goes for planets as well as stars. Overall, it takes about a hundred million years for a solar system to sort itself out—at around the same time the central star is finally fusing hydrogen for energy, the suite of planets it now finds around it are generally in it for the long haul.

So many growing pains. So much blood and sacrifice required to build a family of planets. Cherish every world you come across, dear traveler, for it has seen far worse in its long and tortured life than you could possibly imagine. Pay your respects and move on.

And whatever you do, only visit a solar system if it is mature enough to welcome the occasional visitor.

Stellar-Mass Black Holes

Black holes?
Black hearts!
They'll eat your bones first,
and save your soul for later.
—Rhyme of the Ancient Astronomer

Admit it, you skipped right to here without reading any of the previous chapters.

It's fine, I'll allow it. But this is going to be one of the biggest headaches you're likely to encounter, partially because of the insurmountable danger that black holes pose to the incautious traveler, and in part because black holes stretch the bounds of known physics. We know they exist, which was up for considerable debate back in the day, and we know that they're dangerous. We have a somewhat concrete understanding of the physics outside them—which I will highlight to our gruesome amusement here in this chapter—but they present so many fundamental paradoxes and riddles that when it comes to understanding them, your mind quickly becomes engulfed in unending and eternal darkness.

Black holes. The eaters. Unstoppable, insatiable beasts, lurking in the depths of the galaxies, waiting in eternal patience for their next victim. Some make themselves known through the vociferous consumption of a nearby star or wandering cloud of gas. Most are simply silent, rogue, quiet, and implacably menacing. Many brave explorers have met their horrible and disgusting ends after falling into the clutches of one of these horrid pits. Their lives and souls lost to the rest of the universe for all eternity.

Learn from their mistakes.

Please.

While everything in the universe is trying to pull you in with gravity—except in extreme cases, that's all that gravity knows how to do—all but black holes will let you back out, if you try hard enough, that is.

Bigger and more massive objects require more velocity to escape their gravitational embrace. This conveniently named quantity, "escape velocity," depends on the mass of an object and your distance from its center. More massive = harder to escape. Closer to center = also harder to escape. This works because the definition of escape velocity is based on energy. If you were to launch yourself from the surface of an object with a certain speed, the gravity of that object will want to slow you down. You'll still go *away*, just ever slower as you travel, because the gravity won't stop pulling on you, but will continually get weaker. If you launch yourself with the correctly precise escape velocity, by the time your speed drops to zero, you'll be exactly infinity far away.

That sounds weird, but that's just the way physicists like to construct things: the escape velocity is the velocity needed such that your kinetic energy (which is related to your speed) is equal to the potential energy (which is related to your distance) of your current position.

For example, on the surface of terrestrial planets like the Earth, the escape velocity is quite large, requiring directional blowing-things-up (aka "rockets") to overcome their gravitational pull. But if you're already in orbit, it's relatively easy to continue on your outward trek. Smaller objects, like asteroids or comets, need only good hearty shove and away you go. You get the idea.

The thing that makes a black hole so black is that the escape velocity from its surface (and I'll get to what I mean by "surface" later) is greater than the speed of light. "Oh, to escape a black hole I'll just go faster than the speed of light." Don't bother; you can't. Period. Nothing in our universe travels faster than light.

I said, don't even try.

Since not even swift-footed light can escape a black hole, they are by very definition not emitting anything, hence all the blackness. And since escape velocity depends on both mass and distance, black holes get their characteristic blackness by being both incredibly massive and also really small for their mass.

To give you an example, a black hole a few times the mass of the sun will only be a few scant miles across.

The surface, the place where the escape velocity is exactly equal to the speed of light, is called the *event horizon*. The word "horizon" takes its place in the name because it's an edge, a boundary. If you're standing on a planet, the limit of what you can see on the surface is the horizon. Beyond that is a new world, the complete unknown. And so the horizon of a black hole marks the entry point into a new, unknown, and potentially unknowable world.

As for the word "event?" Well, I guess that's because that's where fun things happen. You know, like a party. Where you die.

Like I said, I'll talk about what (we think) happens when you cross event horizons in more detail later in another chapter—I would hate to overwhelm you here when there's already so much delicious danger to unpack—but for now just consider it shorthand for "don't cross this line, or else."

It was Schwarzschild, Karl Schwarzschild, who first identified the event horizon. He first saw the monster hiding inside Einstein's equations of general relativity.[1] And for the longest time, we didn't even believe they existed—that's just how vile they seemed.

The way we understand gravity through general relativity is easy to summarize. In practice, it's lot of complicated math, but in words it's not so bad. General relativity (GR) splits the universe into two camps: dancers and the dance floor. The dancers are all the *stuff*: planets, people, particles. Anything with mass, momentum, or energy. The dance floor is spacetime. Notice that it's one word: spacetime. Not "space and time." Spacetime. You need to use that word and nothing but that word from now on.[2]

Anyway, the spacetime dance floor is a little weird: it can bend and flex and warp. A dancer, standing on the floor, will make a dimple. A fat dancer will make a bigger dimple. If you're waltzing nearby that fat dancer, and you get too close, you can't help but stumble into their dimple and into their eager, sweaty arms. So the dancers moving about on the floor create dimples and ripples, and those dimples and ripples affect how the dancers move.

This is the magic of GR in the shortest way I can possibly describe. What we call the force of gravity is a result of the machinations of the geometry of spacetime.

That's nice, but what does that have to do with black hole and event horizons? I'm glad you asked.

The math of GR is a machine; a tool. It doesn't really tell you anything about the universe itself, but tells you how to solve your own particular gravitational problems. We can't just lean on Einstein forever; we're going to have to sweat a little bit for ourselves. In the case of GR, we need to input the situation that we want to try to understand, and then and only then will Albert tell us how the gravity should behave.

So as soon as Einstein published the raw material for GR in 1915,[3] physicists around the world began applying it to any situation they could dream up. And the first, and simplest solution anybody could think of was this: what is the gravity like around a single, spherical ball of matter? And Schwarzschild was the first one to put pen to paper to explore what dear old Uncle Albert had to say about this completely benign, innocuous, bland, everyday scenario.

Surprise! Black holes.

But of course, it wasn't obvious at the time.

Einstein's equations were perfectly up to the task of describe the gravity of a simple ball of matter—I mean, if it couldn't, then Einstein would have to go back to the chalkboard. For example, Schwarzschild was able to perfectly apply Einstein's machine to the gravity of the sun, which by all accounts is a large ball of matter, and use that to understand how the planets behave in their orbits.

But for complete lack of a better term, in Schwarzschild's solution for the gravity due to large balls of matter, there was a . . . funny thing.

There was a place where the solution went haywire, where nothing made sense. This place was at a particular radius from the center of the large ball of matter, and it depended on how big the ball of matter was.

Honestly, it was just strange. The proper term for it is "coordinate singularity," which doesn't really help our understanding, so we'll just leave that bit of jargon there and keep exploring.

This particular distance, this radius, quickly got a new name, the Schwarzschild radius. This radius is where Einstein's math didn't make any sense. But this radius was very small. For the sun, it was only a few miles across. And the math behaved perfectly coherently *outside* this radius, so everyone just shrugged their shoulders and figured that this was some weird, but otherwise inconsequential, quirk in Einstein's equations.

Yeah, the Schwarzschild radius was weird, but I guess it didn't really matter. Right, guys?

But then the question immediately popped up from some smart aleck or another: what would happen—just go on this ride with me for a little bit—if we squished a ball of matter to be smaller than its Schwarzschild radius? Would it be OK? Would something funny happen?

The math of GR tells you exactly what would happen: complete, total, irreversible, incomprehensible, reprehensible, insensible collapse. *If you, somehow, squished down a ball of matter below its own Schwarz-schild radius, gravity would simply win*, with nothing left to stop it, and everything collapsing down to the center, with the gravity surrounding that—everything past the Schwarzschild radius, which was once called the *coordinate singularity* but soon acquired the new name of *event horizon*—so overwhelming that nothing, not even light itself, could escape.

You would make a black hole.

✪

Black holes can technically form from anything, anywhere. All you need to do is squeeze enough stuff together in a small enough space, and once the escape velocity is greater than the speed of light (or in other words, when an object as shrunk beneath its own Schwarzschild radius), voilà: black hole. "What happens *inside* the event horizon?" you may righteously ask. Not yet, grasshopper, not yet. We'll get to that in another chapter, because frankly, if you find yourself in a situation where you get to learn firsthand about the interior of an event horizon . . . just . . . well . . . it was nice knowing you.

I should note that Schwarzschild did not personally experience any black holes in his lifetime; he figured all this out from the safety of math.

You can take the sun and squeeze it into a black hole about 5 kilometers across—that's its Schwarzschild radius. You can take the Earth and squeeze it into a black hole about the size of a peanut. If you're allergic to peanuts, it doesn't matter, because you just made a black hole and you have more important problems right now.

You can even take yourself, I suppose, and squeeze you into a black hole about the size of an atomic nucleus. Difficult, I know, but there are tales of a school of mystics who devote their lives to attempting it. It's a strange universe.

Yeah, that's pretty weird alright. And for the longest time folks just assumed that this was just a weird random quirk of GR—I mean, Einstein himself had more than a couple quirks himself so it should come as no surprise that his brainchild has a few eccentricities—but didn't represent anything actually physical.

Sure, these "black holes" appear as an artifact, but nature doesn't actually *make* these. Right?

But gravity is generally great at sucking, and frankly would much rather have everyone cuddle up in a great big galactic group hug, so really the question isn't "Can black holes form?" but rather "How come black holes aren't as common as mosquitos in July? What's *stopping* them?"

You yourself aren't a black hole because you're not as fat as you think you are when you look in the mirror, and gravity is kind of weak. (Actually, it's by far the weakest of the forces. Seriously, even if it were a billion, billion, billion, billion times stronger than it is, it would *still* be the weakest force. Why is it so weak? Good question, glad you asked. We don't know.[4])

Also, bones. You have bones. A skeleton. Your skeleton keeps your body from collapsing in on itself.

Earth isn't a black hole because it's made of rock, and rock is strong. Well, stronger than gravity, which is important here. The fundamental force here is electromagnetism: the electrons in atoms repel electrons in neighboring atoms, and that force is much, much stronger than the tendency of gravity to pull everything together. This force prevents you from falling through your chair, and it prevents the Earth from collapsing in on itself.

So your body and the Earth are in no danger of becoming a black hole any time soon, because other forces easily overwhelm the general suckiness of gravity, which is a relief.

But what about stars? What about the sun?

The sun isn't made of rock, but it also isn't a black hole, since it's in a constant state of blowing up. The nuclear furnaces in the core provide enormous amounts of energy and pressure, heating up the outer layers of the sun and preventing gravity from winning. And it's beautifully self-regulating, too: add a little bit of mass, and the fusion reactions will heat up to provide extra output. If the nuclear reactions died down a bit, the sun would collapse a tiny bit in response, increasing the core temperatures and pressures enough to return to balance.

But eventually the sun, and every single star in the night sky, both seen and unseen, will run out of fuel. Every star will die. And with nothing to fight against gravity—no bones, no rocks—could the bright and brilliant stars of our universe turn into hideous monsters in their deaths?

Short version: yup.

Long version: Even after the sun dies, it's not quite big enough for the Big Collapse. We'll get to this later, but for now let's just say that you can only pack electrons or neutrons in so tightly together before they just won't budge without enormous forces. Thankfully, even after fusion gives up, there are enough other forces laying around to prop up the dead remnants of sunlike stars, prevent total collapse and a life of blackness.

No, to squeeze stuff small enough to form a black hole, to make enough self-gravity that nothing can stop you, to allow you to finally fold inward beneath your own Schwarzschild radius, you need to start with *a lot* of stuff. At least eight to ten times more stuff than the sun. Then and only then is there enough gravity to overwhelm any other forces and create a black hole. Even then it's tricky: the biggest stars tend to violently explode when they die (to be delightfully explored later), and most of their mass gets blown off into deep space. So you need a really, *really* big star; one that has enough stuff left over after the explosion to give gravity enough to work with and create a black hole.

★

You never know where another black hole may hide. In fact, they're so well hidden that for decades after Schwarzschild's revelation, we just flat-out assumed that something—*anything*—would prevent the Ultimate Collapse. We had figured out that stars can die, and we figured out that this process can result in some, suffice to say it, *extreme* gravitational scenarios.

But still, black hole? I mean, come. On. People. You really want me to believe that nature makes objects *literally made of gravity*?

Yeah, right.

And as for the theorists? It was starting to get a little heated out there. Some were absolutely convinced that something would prevent the most massive stars from turning into black holes; that there was some unknown process or new physics to stop all this nonsense. Others, though, were starting to get convinced. There was nothing that they could see to *rule out* black holes. And if nature doesn't expressly forbid something, then it *must* exist.

And so, how to find them? How to test this waking nightmare of Schwarzschild?

How can they possibly be found, if they are so dark and menacing? Did I mention that black holes are small? Oh, I didn't? Well, you can't expect me to keep track of all this. A typical black hole will be a few times more massive than the sun, with a Schwarzschild radius of only a few miles. Compared to everything else in the universe, that's petite. They can get much bigger, of course, but those are far, far rarer, and we'll worry about those when the time comes. You're still using training wheels here.

And black holes are . . . kind of black. A small black hole in front of the emptiness of space is black-on-black. Not a lot of contrast.

How can some surface-dwelling astronomer or some happy-go-lucky explorer find them, out there in the depths?

I hope you're up for a challenge.

Let's say that in your travels you come upon a common sight: a binary star system. Despite these two stars being born together—making them nuclear twins—they are not equal. The life of a star is determined by its mass: the bigger the star, the stronger the gravity, the more ferocious the nuclear reactions, the faster it runs through its fuel, and the shorter its life.

So you can have two stars, born together, but one of the siblings has burned its candle at both ends, and is already dead and gone. To become what? A black hole? A neutron star? Something else? Nothing at all?

The remaining star will reveal the fate of its sibling. Locked to its dying companion with invisible chains of gravity, it was forced to watch the whole macabre dance play out. But soon enough, it too will near the end of its life, swelling to become a giant red monster of a near-dead star.

But this aging beast is still in orbit around in its long-dead companion, whatever form it's taken. The outermost layers of the red giant swollen and tenuous, some of that atmosphere will spill onto the companion, the gas swirling and spinning as it's pulled onto the other object, whether it be some stellar remnant or some darker terror.

From a distance, a black hole is almost indistinguishable from some other possibility. Shrouded in a disk of gas, its true nature is hidden. But as it falls in, the gas itself blazes with intense radiation, heating up from gravity and friction as it crams down onto the surface of the mysterious companion.

And so here you are, a cautious and prudent traveler, trying to assess the danger of any potential threat in this binary system. I suggest that you always assume you are dealing with a black hole, unless you can prove otherwise. Remember: neutron stars and other remnants may be sources of extreme gravity and radiation, but at least you can leave when the party's over.

The best way to determine if you are dealing with a black hole is to determine its mass and size. These two combined can be used to compute the escape velocity. If it's less than the speed of light, you're good to go—literally. The mass is easy: just measure the orbital period of the larger red giant companion and use Kepler's laws to compute the mass. You did bring a calculator, right?

The size is a little bit trickier, because of the dramatic mystery-creating shroud of gas and all. But you can figure out a *maximum possible* size. Look into the center of the cloud of gas. See anything? If no, look more toward the center. Keep going. The central object—black hole or something else—has to be smaller than that. Because otherwise you would've seen it by now.

Look, astronomers on Earth can do this from thousands of light-years away; it's not hard.

And here's where I have to tell you that it looks like Mother Nature is as crazy as we feared. Using observations of systems exactly like this, Earth-based astronomers found their first black hole candidate in the constellation Cygnus. They found a bright source of X-rays (dubbed Cygnus X-1, of course, because it was the first and brightest source of X-rays in the deep sky[5]), and after considerable and heroic effort, figured out the mass and size of the central object, and . . . well, it's too small and too massive to fit any other description.

Whatever's going on down there in the Cygnus X-1 system, it looks and acts like a black hole.

And that black hole is not alone.

This method applies, of course, only to lucky binary systems, where you can use a convenient orbiting neighbor to compute the mass and size of the hidden companion, revealing its true nature. What about lone wolves? Black holes kicked out of their original systems—probably because they ate all the snacks—and roaming free through the galaxy?

I'm pretty sure you can spot a rogue white dwarf or neutron star (or even planet or brown dwarf), because they tend to be glowing in at least some band of radiation, making them in some sense visible to any variety or combination of instruments you may have with you.

But like I said before and should be abundantly clear by now, black holes aren't just a little bit black—they're the *definition* of blackness. Complete and total absence of light. Lucky for you, you can still spot a rogue black hole, but in a very zen way you can't look for where the black hole itself is—you have to look where it *isn't*. For this trick to work you have to remember the lessons that Albert tried to teach you: massive objects bend spacetime, and that bending of spacetime tells all things how to move.

All things: including light itself.

OK, quick lesson. Your survival may be at stake. Gravity bends the path of light. "But light doesn't weigh anything, and I thought gravity only pulls things with mass." You thought wrong. Where do you read things like this? Gravity doesn't care what it pulls; it will even pull you and your head full of wrong. Gravity acts on anything with mass, momentum, or energy. Light has momentum, ergo gravity can pull it.

Now, where was I? Oh yes, gravity bends the path of light, and heavy things like black holes can act like a lens: bits of light that would've just grazed the surface of the black hole get bent and refocused into a new direction. If you look carefully at a field of stars, then every once in a random while a random black hole will cross your line of sight to one of those distant stars, and the light from that star will get momentarily twisted, distorted, bent, twinkled, tweaked, and maybe even twerked.

Once you notice that unusual twinkling (Remember, once you're above a planetary atmosphere the stars should settle into tiny, painful points of light. If you sense any scintillation then something's amiss.), you can use the power of general relativity to figure out the mass and size of the offending object from the amount of starlight tweak.[6]

Of course, there are thousands of other things that can affect background starlight like that. Sometimes stars just flare and surge and vary for their own closely held reasons. Sometimes an orbiting pair can be confused for a single, oddly behaving object. To run a real black hole detection search, you'll need sophisticated computer algorithms and some serious hardware to get the job done. But hey, there's a reason we don't let any old idiot into space.

✪

So far, I've only talked about relatively small black holes—around a few times more massive than the sun. These are the remnants of the very largest stars. There are of course larger black holes (much larger), and we'll get to those beasts soon enough. Are there smaller ones? Maybe, depends.

I mentioned tiny black holes briefly when talking about cosmic rays and all the mischief those microscopic particles can get up to. Overall they're harmless (or at least, no more harmless than any other fast-moving subatomic object), and in some varieties of fancy theoretical models of fundamental particles and forces, occasionally certain exotic interactions can make a black hole. It would be a teensy tiny one—microscopic, actually, not any more massive than any other exotic particle. Once created and left to its own devices, that black hole won't bother anybody, but by chance it might nibble a stray electron or neutrino. It grows, increasing the strength of its gravitational pull. Bit by bit, nibble by nibble, it could get massive enough to be Taken Seriously.

Except for the whole evaporation thing.

Sigh. You know all that poetic language I devoted to describing the absolute pure blackness of black holes? Right, about that. It turns out that black holes aren't, after all, 100 percent completely, totally black.

Hang with me here a minute.

According to our current understanding of quantum mechanics in strong gravitational environments—which, to be fair, can at best be described as "sketchy"—black holes can evaporate. It's a process called Hawking radiation, and it just so happened to be discovered by Stephen Hawking, hence the name.[7]

It's not a straightforward process at all, even for fans of quantum mechanics (all three of them in the universe). In the standard telling of Hawking radiation, someday a random particle appears in the vacuum of spacetime with a *poof* of quantum energies, say, an electron. But it's not alone, due to various arcane quantum mechanical rules, it must be joined in its brief existence by its antiparticle (which is a particle with the exact same properties but opposite electric charge), in this case, a positron.

Yes, particles can appear randomly for short enough times, because that's the way our universe works.

In most cases, these particles, newly created from nothing, usually find each other and explode, giving back their borrowed energy to the vacuum of spacetime. No trace of their brief existence remaining. All this happens in the briefest possible amount of time. No harm, no foul.

But if this happens near the event horizon of a black hole, one partner may be unlucky and sucked in, while the other gets off scot-free.

Particles emitted from the surface of a black hole? Looks like radiation to me. And where do they get that energy? From the black hole itself. There's simply no other option. The universe found itself host to a particle that sprung from the vacuum—*somebody's* got to pay for it, and the bill lands squarely in the black hole's mailbox. In return, the black hole has to pay, and it does so by losing a little bit of mass.

I should mention, as the trustworthy travel guide that I am, that I'm not entirely comfortable with this approach to describing Hawking radiation. It's not exactly *wrong*, per se, but it is pretty far removed from the mathematics of his original paper. And you know how much I love math.

But a straightforward rundown of Hawking's math isn't exactly illuminating (ha!) either. It's less about particles popping in and out of the vacuum (which is a very real thing), but about the fundamental quantum fields of spacetime themselves. Remember *those*?

There are a bunch of quantum fields all doing their buzzing, vibrating thing at all times, even in vacuum. Minding their own business. Then, suddenly, a star dies and a black hole forms! Some of those quantum fields will get trapped midvibration, hanging out near the event horizon for ages upon ages. But eventually, they will be released (the ones that formed *inside* the event horizon are trapped forever; the ones near that one-way boundary are merely trapped for a painfully long time).

From our perspective, watching the black hole millions or billions of years after its initial formation, the vibration of the quantum field appears to escape the black hole, and it turns out that the black hole wasn't quite as massive as we had thought just a moment ago.

In Hawking's math, this process of radiation is less about particles getting trapped inside the event horizon, and more about a complicated, intricate dance between the event of the birth of a black hole and the quantum fields that soak our universe, making a black hole appear larger than it really is, and giving them a finite lifetime. And in the process, making them glow.

But the upshot is that no matter how you slice it, despite always being hungry and having an insatiable appetite to grow ever larger, beings of pure gravity that they are, if left completely alone black holes will actually lose weight.

This process is so slow it barely even counts. For a typical black hole a few times the mass of the sun (Estimates are a bit rough here. In the Milky Way there's somewhere between 10 million and a billion black holes. Like I said, rough.) the total energetic output of Hawking radiation is somewhere in the ballpark of a single photon every year.

That's . . . slow.

It takes well over 10^{100} years for a black hole to finally evaporate. It's just not something we ever need to worry about on human timescales.

But the smaller the black hole, the faster this process runs. So any tiny microscopic black holes go poof before they have a chance to grow. And even if they *did* survive long enough, they're starting off so small that they take a *really* long time to grow. Getting an electron to hit a microscopic black hole would be like trying to hit a baseball with another baseball . . . from a thousand miles away. So there could be microscopic black holes running around all over the place, but you wouldn't ever notice them because they're so dang tiny and inconsequential.

Is there anything else that can form a black hole? As far as we can tell, the only known mechanism is the death of massive stars. Some theorists had wondered if the very early, very young, and very wild-and-crazy universe could've directly produced some black holes (imagine, for example, a piece of spacetime spontaneously forming a black hole *simply because it felt like it*), but observations of the early universe haven't borne that out.

There might be some funny business going on in the first few hundred million years of the existence of the universe, when the stars and galaxies were first coming online, where enough material could have collapsed to form black holes without going through the whole hey-guys-let's-be-a-star first stage. That may be a possible source of the giant black holes in the universe, which I keep teasing but refuse to talk about until they have enough space (pun intended) to fill their own chapter.

✪

I don't think I've properly expressed yet just how dangerous black holes are. I'll sum it up with a statement in all capital letters, so you get the impression that I'm yelling at you: AVOID BLACK HOLES AT ALL COSTS.

Yes, the event horizons are one-way tickets to oblivion, that much should be obvious. But don't go chasing dragon tails—don't even get close to them.

Let's say you get the wonderfully stupid idea to investigate the local regions of a black hole. Maybe you want to study gravity at a more fundamental level. Maybe you want to investigate extreme plasma physics as gas gets pulled in from a neighboring star. Maybe you have a perverse death wish.

To show you what would await you should you be foolhardy enough to venture close to a black hole, I'll introduce two explorers: Alice and Bob. Alice is going to attempt to travel close to the black hole and will die a wonderfully comical death. Bob will be smart and observe from a distance. Don't worry: Bob will get his turn in a later chapter to die his own peculiar death. But for now, he's safe.

From a far enough distance, there isn't anything special about a black hole—gravitationally at least. It's still a menacing and ominous hole in the universe, mind you. But you can orbit it just like any other object. Gravity is pretty cool like that: from far enough away, two objects will interact with each other as if all their mass was in a single point at the center of each object. You don't need to care about odd lumps or bulging equators when computing the effects of gravity. Just total mass and distance.

Orbiting a black hole the mass of two suns would be exactly like orbiting a star with the mass of two suns. I mean, there wouldn't be any heat or light or warmth or sunshine or anything else to make life worth living, but hey, you could orbit it.

Our explorers, Alice and Bob, will stay perfectly safe as long as they keep far from the black hole. How far? It depends on the exact mass, but "no distance is far enough" is fair advice. Although if you want to be technical, at least ten times the Schwarzschild radius of the black hole will suffice as a general rule of thumb.

It's only when Alice decides to go on a little adventure do things get . . . interesting. The strength of gravity that you feel from an object depends on how far away from it you are. Really close to something? Strong gravitational pull. Really far from something? Weak gravitational pull. In fact, the effect is *squared*, so if you double your distance from something, the gravitational pull gets cut down to a quarter of its previous strength.

If you're reading this on Earth, you feel a nice strong pull from the Earth, but you hopefully don't feel one from Jupiter. Even though Jupiter is so much more massive than the Earth, it's really far away from you, so who cares?

This means that your feet are slightly heavier than your head. And not just because you're brainless. Your feet are just a little bit closer to the Earth than your head. Not by much, just a meter or two, but it's not zero. Your feet are closer to the ground, so they get a slightly stronger gravitational pull, hence the increased weight.

So, while your whole body is being pulled by the Earth, your feet get a slightly stronger pull than your head. It works in the opposite direction too: your head is a little bit further from the Earth, so it's actually ever-so-slightly floating away from your body. Two forces, equal and opposite: your feet getting an extra tug, and your head getting an extra lift.

Remember that bit about gravity only caring about the centers of objects? Yeah, I lied. Well, not *technically* lied. What I said was perfectly true, for large enough distances between objects. But now we're getting intimate, aren't we?

This is called a *tidal* effect, since it's responsible for the . . . wait for it . . . tides. One side of the Earth is slightly closer to the Moon than everything else, so it gets an extra tug, the oceans lifting up trying vainly to reach the moon, and there you go, a tide on that side of the planet. At the exact same time, the ocean on the opposite end of the Earth from the moon gets feeling a little left behind, and tries to float away—a tide appears there, too.[8]

Another way to think about this tide on the opposite side of the Earth is that the whole entire Earth is getting slightly pulled toward the moon, *except* for that bit of ocean that's too far away. Poor ocean.

The sun does the same thing to Earth's oceans, too, making tides on our home planet a little tricky to figure out, which is why surfers have such complicated lives.

But these tidal effects don't really affect your body. At least I hope not. That's because the difference between your head and feet—a meter or two—is small compared to the distance to the center of the Earth—millions of meters. So the overall tidal stretching that you might experience will be incredibly small. Same for the ocean tides: it takes a whole moon's worth of stuff just to make a relatively itty-bitty tide.

But you're not near a black hole, and Alice is now approaching one. Since black holes have so much mass—over a sun's worth of mass—compacted into such a small space, as Alice nears the black hole the tidal effects become larger and larger: in other words, the difference in gravitational pulls between her feet and her head goes from unnoticeable, to mildly irritating, to panic-inducing.

Her entire body falls to the black hole—probably uncontrollably because she didn't even bring a jet pack—because that's exactly what gravity loves to do. But because of these tidal forces, her feet fall a little bit faster and her head falls a little bit slower. It's like two hands on either end, tugging, tugging, tugging.

But wait, there's more! When we think "gravity" we generally think "down." Good enough, for planet-bound folks. But remember, gravity points to the *center*, and the exact, geometric center of the Earth isn't 100 percent down, but down and a little bit *in*. Imagine two arrows stuck to your waist, pointing toward the center of the Earth. They'll be pointing *almost* perfectly down, but not quite. They will be pointing just a hair's width inward.

It's not a big enough difference to matter on the Earth, but Alice isn't on Earth, and she's starting to feel a little crampy, and not because of her lunch. Hold your fists at your waist. I don't care if you're reading this on an airplane, do it. Those points on your waist don't just feel *down* gravity, but *down and a little bit in* gravity. Alice, near the black holes, feels an extreme version of this. At the same time the tidal effects are stretching her out, this extra gravitational direction is literally squeezing her midsection like a tube of human toothpaste.

The combined effects—pulling and squeezing—mean that Alice isn't going to last long. Surprisingly, gravity being gravity and the sizes of black holes being the sizes of black holes, if you're freely falling toward one of these beasts (where "freely falling" means exactly that—you're not fighting your movements in any direction), the gravitational effects will become extreme enough to rip apart just about any atomic bond about one-tenth of a second before you strike the black hole.

The original Earth-based scientists who discovered this effect must have been in a lighthearted mood that day, since the totally legit, 100 percent scientific technical term for this effect is . . . *spaghettification*.[9] You can find that word in your friendly neighborhood astrophysical journal. Spaghettification. *Come la pasta*, as the Italians would correctly observe.

★

But wait, there's even more good news. And by good news I mean horrible news, for Alice. Well, what's left of her. She's been atomized, the gravitational forces overwhelming the bonds that hold her together. She's now just a thin stream of molecules, and eventually atoms, and eventually subatomic particles.

The tidal effects I just talked about are all "normal" gravity. It's just good ole fashioned Newton's equations, but in a really extreme environment. But closer to the black hole, Newton isn't good enough and we have to switch over to super-Newton, aka Einstein's general theory of relativity, which at the time of writing is our current best-guess story of how gravity works.

And oh, what a story it is to tell.

You see, black holes are typically spinning. That's because they're born from parent stars, and stars spin. As the star collapses to form a black hole, its spin only increases to frightening speeds. And if they're feeding? Oh boy. The gas falling in is orbiting, so as the gas enters the black hole's hungry mouth, it gives it a nudge, like pushing a merry-go-round of doom.

OK, so what, a spinning black hole. What's the big deal? Well, according to general relativity, massive, spinning objects can drag spacetime with them. I know, it seems weird. Let me break this down for you.

Remember the analogy with the dancers on the dance floor earlier? Now imagine a spinning dancer, pushing off their toes over and over again, going as fast as they can. If they're heavy enough and fast enough, they'll start to tug at the flexible dance floor underneath them. The floor itself will start to wind up around the dancers' feet. In other words, a spinning dancer can drag the floor with them.

A spinning black hole can drag spacetime with it. Get it now?

As what remains of Alice's feet near the event horizon, they get caught up in a region called the *ergosphere*, a region of space surrounding the spinning black hole where the dance floor is twisting up. The atoms that were once a part of Alice's feet will get spun around and around the black hole, trapped in the ergosphere, unable to resist the twisting, before finally falling to their final release in the event horizon.

A strand of spaghetti, wound up on the end of a fork.

Come la pasta, indeed.

★

Now I won't talk about what becomes of Alice's atoms after they pass through the event horizon, since it's not like you'll be there to check. We'll save the tales of black hole interiors for another chapter, and then I'll tell you the truth about singularities.

Bob, after saying his final farewell to the now-spaghettified Alice, continues on his way to other—perhaps safer—parts of the galaxy. He, like you, now knows what to look out for. Smaller black holes—called *stellar mass* by sophisticated planet-bound astronomers—are generally sprinkled here and there throughout a galaxy, some bound up in binary systems, most freely floating and wandering the depths.

Since they form from the deaths of massive stars, you have a heightened danger of encountering one in young stellar systems. Here the first generation of stars has already gone out, either sputtering on as neutron stars or encased behind event horizons. Occasionally, of course, black holes can be kicked out of their original systems. Going rogue, lone-wolf style, not bound to any system of any laws—except the law of gravity, that is. You never know when you might encounter one on your journeys.[10]

Since they are so small, black holes are relatively easy to avoid—if you know what to look for. If you come across a binary system and are unsure if a companion is a dense stellar remnant or a black hole, assume it's a black hole and get yourself somewhere else. It's not worth it, unless you want me to just douse you in marinara sauce right now.

But even solitary black holes may still be feeding, sipping away at the remnants of their host systems. The surrounding gas, falling to its doom, will heat up, emitting X-rays and even gamma rays before vanishing behind the event horizon. If you see a compact source with extraordinarily high radiation levels, back off.

It's the quiet ones that are most dangerous. With nothing to feed on, they are lean and hungry, desperate for another meal before Hawking radiation finishes them off. That also means that there are no visible signs of the presence. Keep a careful watch on your instruments, looking for the subtle twinkling of starlight by otherwise-hidden monsters.

Forever more, I want you to associate these two phrases in your brain: "black holes" and "never mind."

Planetary Nebulae

You put on a nice show
A neon sign—fun is here
But truth is grim
A rusting place, best forgotten.
　　　　　—Rhyme of the Ancient Astronomer

We've come to an interesting place in our journey through the galaxy. We've broken out of our home solar system after dodging rogue asteroids, evading circuitry-frying coronal mass ejections from the sun, and simply accepting the flood of tiny cosmic rays constantly bombarding our delicate flesh.

Once we reached interstellar distances, we've seen stars born from clouds of vicious turbulence, and we encountered our first truly exotic creatures of this everlasting night that we call outer space: the black holes.

Those black holes are tombstones.

Markers. Memories of what once was. Almost-forgotten remnants of the past. The black holes of our galaxy used to be stars, shining with heat and light and warmth. Dead, gone now, generations ago. Their fusion finished, their hydrogen depleted, their spirit withered.

The stars of our universe will die, one by one. And those deaths are, to a number, nasty.

Many of the hazards that we're about to explore and explain will come from the variety of ways that stars can end their lives, and turn into other, far less pleasant, things.

Needless to say, it won't be pretty.

And when it comes to stars, everybody wants to go out with a bang. Make a big deal out of it. Sadly, not everybody can shine as brightly, as intensely, as ferociously as a supernova. Don't fret, eager explorer, we'll get to supernovas and other overly powerful explosions soon enough. It's best we start small, with weaker explosions and their consequences, and work out way up to the big leagues. Wouldn't want to get ahead of ourselves.

The Earth's own sun will die someday. Best to accept that fact now, deal with it, internalize it. When the moment comes you don't want to be caught off guard, your eggs half boiled, your grilled cheese sandwich only toasted on one side. We'll use the sun as a textbook lesson, so you won't be caught unawares in an unfamiliar system, so you won't pick a star to call home that will go giant on you in a thousand years.

All stars die. Some, the very largest, go out in a tremendous flash of energy, turning themselves inside out to light up the universe. Others, the smallest, slowly fade, never quite sputtering out, never making a scene, spending a trillion years tending to a weak fire.

The middle ones, like the sun, have the most miserable fates. Before they finally die, they become red and bloated, spewing out their innards through the local system. Spasm after spasm, they slowly lose themselves, leaving only a faint dying heart behind.

It's in these later years when they're most dangerous, when their violence overwhelms them, reducing any hapless inner planets to cinders.

Every year, old stars fade while new ones light up. A continuous cycle in the galaxy. Beautiful and poetic, really, except for the fact that when these stars go they cause mayhem and destruction for anybody unfortunate enough to live within their influence.

✪

It starts slow, creeping. Inch by inch, day by day, the sun brightens. You may not notice it over months, or years, or even centuries. But the clock is ticking. Deep in the sun the nuclear furnaces are kept stoked with a fresh supply of hydrogen in the core. As it burns, it leaves behind freshly minted helium. Atom after atom sinks to the deepest regions of the sun. Without enough pressure to fuse, the helium sits, inert, lifeless, its potential unlit.

The helium ash grows, unburnt and—for now—unburnable, pushing aside the hydrogen, making it harder for those little protons to find each other and join the fusion party. But the crushing weight of the sun's own mass continues to bear down, the gravity as relentless and inexorable as always.

The fusion must keep up and fight the gravity, must maintain the balance no matter what, despite the slow and steady accumulation of useless helium in the core.

To keep combating gravity, the core grows hotter, the temperatures climb higher as the hydrogen fantastically tries to find its brethren in the chaos, fuse, and release the needed energy to keep catastrophe at bay. And as the core grows hotter, the rest of the sun grows larger and brighter in response. Eons ago, the sun was dimmer, smaller, cooler. Eons from now, the sun will be brighter, larger, hotter.

The dinosaurs knew a smaller, dimmer star than we do, and the first life to swim through our oceans billions of years ago would look on in fright and wonder at the blazing monstrosity crossing our sky that we blithely call the sun.[1]

Indeed, we live in a surprisingly lucky age, where the moon is about four hundred times smaller than the sun, but the sun is about four hundred times further away. They are in balance, both taking up an equal amount of our sky, giving the people of Earth a wondrous total eclipse every few years. Soon enough, the sun will be too large and too bright ever to be eclipsed again.

In a process only measured by the millions of years, the sun will cook the Earth and any inhabitants in the inner solar system.

Nothing can stop it. Earthbound life has had almost four billion years to flourish under peaceful conditions, with liquid water to splash around in and a thick atmosphere to filter out almost all the cosmic rays. But within a few short hundred million years—a blink of the cosmic eye—its end will come.

You may have heard that the sun has another four to seven billion years before it finally extinguishes. You're right, and we'll get to that in a bit, but that's not what matters. What matters is the capability of a planet to support unshielded, unprotected life. The Habitable Zone, the ring around each star marking the conditions right for life—not too cold to freeze water, not too hot to boil it away—is continually shifting ever outward.

Every star shares the same fate: a favored planet one day will be a ruined husk the next. A forgotten icy wasteland on the outskirts one day will having flowing rivers and great oceans the next.

Venus has already suffered its grim doom, choking on its own poisonous atmosphere, a traitor sun to blame. The Earth is next, sometime in the next few hundred million years. You think global warming is bad? How about global incineration?

The same process that strangled Venus will come for her sister next in only a few hundred million years.[2] As the sun ages and heats, the inner edge of the Habitable Zone creeps closer. At first it's not much—a little extra water vapor in the atmosphere, no big deal. But that water vapor will trap more heat, which will pull more precious liquid out of the oceans, which will trap more heat, and so on and so on in a deadly dance. Once dry and desiccated, the great churn of the continents, which reshaped and resculpted our surface for billions of years, will grind to a halt. The carbon dioxide locked up in our soil will vent, piling up.

If life is able to cling to some meager existence in some crack or crevice, it won't last long. Soon—all too soon, really, as humans happened to arise in the last days of life on our home planet—our solar system will host a new Venus, perhaps bright enough to shine as the morning star as viewed from another world, but never again to be a refuge for life.

And Venus herself? As dead and barren as Mercury.

The sun will set on a lifeless, hellish Earth, but maybe now Mars will have a chance to relive its glory days. It once hosted liquid water, and maybe then, hundreds of millions of years from now, it may find itself clement once more.

If you must choose to settle permanently on a planet, you must carefully note the age of the star. Will your planet be comfortable for billions of years, or just a few short millennia? A new planet-bound civilization is a dear thing: choose wisely.

Thankfully, astronomers know what's going on, and can roughly ballpark the age of a star just by looking at it. The first to do it were the pair of Ejnar Hertzsprung and Henry Norris Russell, who devised what we now call the Hertzsprung-Russell diagram, or *H-R diagram* for short because who the heck has the time to keep saying "Hertzsprung-Russell?"

They found that stars follow very specific paths as they age, always getting brighter and hotter with time. And the more massive stars evolve much faster than the smaller ones (which makes sense: with more mass, there's more gravity, which means the fusion rates go that much faster, and the whole game plays out in fast-forward). So by looking at a star and measuring its mass, you can pinpoint where on the evolutionary track it is, how old it is, and how much time you've got left until it gets worse.

There are ways to avoid disaster. Of course, all the ways are hideously expensive and are likely to fail, leading to even more certain doom, but they're worth a shot, if you think your planet is precious enough to be worth saving.

Since this solar advance is so slow and creeping, taking hundreds of millions and billions of years (for reference, our sun is about 30 percent brighter than when it was born, and will eventually reach a bit less than twice its current brightness before getting truly nasty) you have time on your side. As the Habitable Zone shifts, you can shift your planet with it.

For example, you can steal some speed from an outer planet, and give it to the Earth.

A nice big asteroid or smaller moon will do. Just line it up right and give it a little nudge. Aim it for Jupiter. Why not: it's big and it's close. Wrap the asteroid's path around the gas giant. As it loops, it will siphon off just a bit of that great planet's momentum through a brief exchange of gravity, slowing Jupiter and bringing it an inch closer to the sun and speeding up the asteroid in return. These kinds of "slingshot" maneuvers are exactly what we use to give our spacecraft a nice velocity boost when they need it the most.

But instead of aiming away from our solar system for the adventures of interstellar travel, loop the asteroid back toward the Earth, and orbit the opposite way, slowing the asteroid back down and speeding up our home world, which nudges that planet further away from the sun.

Repeat the trick, over and over, trading speed from Jupiter to the Earth. It won't be much, of course, a little nudge with every flyby. But you don't need to do anything drastic: slow and steady wins this race.

Carefully initiated, the process can be essentially left on autopilot. As the sun gets warmer over the eons, the Earth backs off, cooling itself against the increasing inferno, staying well within the bounds of the Habitable Zone.

Of course, if you get it wrong, either the asteroid wanders off into deep space (best case scenario) or comes crashing down on Earth (worst case scenario). Hope you double-checked your math.

✪

At least with careful planning and a little luck, you can save a planet from ruin from increased solar brightness. If you can play that game for billions of years, matching one notch of more brightness with one notch of further orbit, you can stay habitable despite the looming threat.

But that trick will only work for so long. (Where "so long" can mean billions of years, but it's still finite and somewhat short compared to the lifetime of the entire universe; I mean, you actually want to explore for a while and get your money's worth, right?) As stars burn through hydrogen, in a stage of their evolution known as the main sequence, everything is pretty stable and predictable. Hotter and bigger with time, yes, but that's about as stable and predictable as things go in our cosmos.

Unfortunately for you, eventually the gas runs out.

The irony is that most of the hydrogen in a star like our sun will go unfused. In order to play the fusion game and turn into helium, the hydrogen has to be buried way down deep in the core where the temperatures are scorching enough and the pressures intense enough. Way out in the upper layers, things are just too thin and too cool for anything nuclear to happen. Just look (not with your eyes, please) at the surface of the sun—just a few measly thousand kelvin! You think any protons are going to smoosh together in those conditions? It might as well be Siberia in the winter for how cold that is.

So the lifetime of a star like our sun isn't limited by the total amount of hydrogen, but by how much hydrogen is able to get into the core. You may have gallons and gallons of gas in your car, but unless it's *actually in the gas tank* it's not going to get you down the road.

This is one of the reasons that the small red dwarfs are able to live such achingly long times. Not only does their relatively diminutive stature mean that the gravitational pressures in the core are weaker, leading to a somewhat sedate fusion rate, but their interiors convect up and down, churning through their entire atmospheres, continually pulling fresh hydrogen from the topmost edges to the central reactor.[3]

Seriously, if you really want to find a star to set up a long-term retirement community to orbit around, your best bet is something half the size of the sun. As long as you can get used to the cool, red light, you'll find that you'll get to enjoy it for trillions of years, compared to the billions of years of light from a sunlike star.

And the stars bigger than the sun? Yeah, those get their whole own chapter, don't worry.

Anyway, right now we're focused on sunlike stars. You have to hand it to them: they can burn hydrogen and regulate their own fusion reactions for billions of years. We'll take it when we can get it.

But that nasty unburnt helium ash continues to build up in the core. For now, and for billions of years to come, It's not That Big a Deal.

But in about 4–5 billion years, it will become a big deal.

That's when the—ahem—*activity* begins.[4]

Eventually so much helium builds up in the core that the fusion of hydrogen becomes straight-up impossible. And so it stops. Just like that, after billions and billions of years of steady hydrogen-burning in the core, it's just . . . over. Done. With nothing more than a lame shrug, our sun will begin to die.

And now the dark dance begins.

The rest of the star is still squeezing down on the core with all that gravitational might. But with the fusion power unplugged, there's nothing to fight it. And so the helium core compresses and compresses and compresses. As it squeezes in on itself, its temperature rises to millions of degrees. The intensity of the heat from the helium core doesn't do anything to the helium itself (yet), but it does heat up the hydrogen in the layer surrounding it.

And the hydrogen there reignites.

Fusion, again. The lights are back on. But it's different now, strange. Our sun will not be fusing hydrogen in its core as it did successfully for billions of years, but in a thin shell around the hot helium core. With the central power source of the sun now in a shell, this pushes out all the remaining layers, expanding the surface of the sun. And these outer layers, further from the intense core, begin to cool.

I know it's a little tricky, all this inward-and-outward pushing-and-pulling, of contraction and heating and expanding and cooling. In all situations the main goal must be achieved: the balance of gravity crushing the sun into itself versus the force of radiation pressure trying to blow it up. This all stays true in this stage (otherwise the sun would quickly end up as either a black hole or an explosion), but the core and the rest of the sun's atmosphere operate somewhat independently. All the outer layers see is an ever-hotter and ever-larger core burning region, a hot floor that it must constantly tiptoe on lest it get singed. And all the core sees is a heavy stack of books atop its head, trying to crush it down.

All things must balance in a star, and when hydrogen ceases to burn in the core and instead moves into a shell, this is what happens to a star like our sun: it becomes a red giant.

Giant because they're larger than normal, and red because they're . . . red. Really, they're kind of dark orange-ish, but you have to get out from behind an atmosphere to notice that. At least this time old Earth astronomers picked a sensible and accurate jargon word for something out in space, so we'll take it.

The fact that redness is directly tied to something being cooler isn't something that most people appreciate. It's time that changed. Look at the words we use: "red hot," "red in the face." For us, red = heat. But even *hotter* is really blue. So maybe "blue in the face" should be the new "red in the face." Of course, if your face were actually hot enough to glow blue, you would probably incinerate everyone around you, but that's just a minor detail.

Here's why: your body is made of a bunch of atoms and molecules. In fact, you may not realize this, but pretty much everything around you is made of atoms and molecules. I know, crazy, but it's true. All the atoms and all the molecules are wiggling around quite frantically. As they wiggle, wobble, and wibble, they emit light.

When something's hot, the atoms are wiggling faster, because that's kind of the definition of "hot." And when they're wiggling faster, they're emitting light with higher frequencies. When they're cooler, they're moving slower, and they emit light with shorter frequencies.

Your body's temperature is around 100 degrees Fahrenheit, and all your little atoms and molecules are emitting light—infrared light. That's why infrared cameras are so good at seeing people: that's the kind of light we're glowing in. If you heat up a person to a few hundred degrees, their light will shift from infrared up into regular red. Even hotter, they'll go blue. Even *hotter*, they'll go to ultraviolet, X-rays, and even gamma rays. Yikes.

This is called *blackbody radiation*, perhaps one of the most confusingly named phenomena in all of science. Like we saw before, the word "black-body" comes from the experimental device first used to study this way back in Ye Olden Times, and it just sort of stuck. And so the great karmic wheel of astronomy jargon spins around, giving us a nonsense word like "blackbody radiation" for every sensible term like "red giant."

Anyway, because the surface of a red giant star is so far away from the intensity of its central core, it's a lot cooler than a normal main sequence star, hence all the redness (similarly, red dwarf stars are also red because they have a lower surface temperature).

But because red giants are so . . . giant, they have a lot of surface area. Way more surface area than a normal star, with plenty of opportunities to throw light out into the void, so despite being cooler they're actually *brighter*. Which is great news for you: they're easier to spot and easier to avoid.

✪

The core of this new monster becomes twisted and confused. The detailed balancing act that kept the star happy and content for billions of years—the constant interplay between the gravitational crushing and the explosive release of energy—is perverted, and it's all the fault of the helium core.

Once that core reaches a certain mass—for something like our sun the threshold is around the mass of one or two hundred Jupiters—it does something . . . different. It's able to support itself and resist continued expansion not through fusion, but through something far stranger, far weirder, far more . . . quantum. I'm not going to describe it in much detail here—we're much too naive in our explorations of the strange and vicious in the universe. I will, however, tell you that it's called *degeneracy pressure*, and I will warn you that this ability for freakishly dense objects to support themselves without an energy source is going to become a major headache (both in understanding and in danger) later on.

For now, all we need to know is that the helium core stops compressing. It just gets hotter, which makes the shell of burning hydrogen surrounding it increase its fury, which makes the core hotter, which makes the fusion rate higher, and back and forth in a cruel feedback loop.

In other words, the sun spirals out of control.

In just a few hundred billion years after the appearance of degeneracy pressure in the core, the sun swells in brightness, becomes over two thousand times brighter than it is today. Unable to contain the runaway onslaught of energy pumping out of the core, the sun's outer layers bloat to unimaginable proportions. You thought it was a red giant before, when the helium first burrowed to the core and forced the hydrogen fusion into a shell around it?

Well, my little spacefarer, the sun will now truly show you the meaning of *giant*.

If you're living on an inner planet, by the time a star reaches this stage you're pretty much cooked, literally. Once hydrogen fusion ceases in the core, it still takes well over a billion years of shell-based burning before this meltdown phase begins, so it's not like you're going to be taken by surprise, but unless you've got an escape plan, you're just going to sit there and smother.

But the inner planets will suffer the full rage and fury of the dying star. First the atmosphere escapes. Air is a pretty tenuous thing, and if it gets too much energy those lightweight atoms and molecules can get a velocity faster than the escape velocity of their host planet, and away they go, off on their own voyage into deep space.

Next are the oceans. Heat up water in a bucket and soon enough you won't have any water in the bucket. Heat up water on the surface of a planet, and you get the idea.

No air, no water, no life. The warmth-giving friendly sun will become a red raging monster, filling the sky with its hateful gaze.

The innermost planets get really lucky: they get to be stars themselves. Well, not so much individual stars, but more like incinerated leftover planet dust as the star has completely engulfed them. I think that counts, though.

The surface of the new sun becomes a raging torrent: instead of many small convection cells, there are only a few giant ones, literally boiling in response to the ferocity within. And these cells go *deep*, digging up material from the deepest reaches of the inner zones.[5] Hydrogen that hasn't seen open space for 10 billion years is exposed to cold vacuum. Plasma that just a moment ago was undergoing nuclear fusion spills onto the surface, spewing out blasts of cosmic rays and powerful X-ray radiation, far more dangerous to system inhabitants than before. And without a protective atmosphere, any remaining planetary dwellers will be doubly exposed.

Our own sun will expand to grotesque proportions, easily consuming Mercury and Venus, and likely the Earth, too. But that's if our home world is lucky. If it actually survives, it won't be much. Not only will the atmosphere boil away, but also the crust and mantle. Only the iron core—a rump of what was once a proud planet, teeming with life—will be able to survive in orbit around the red monster that it once called the sun.

The bright side to all this is that the outer planets will thaw. The incredibly bright red sun will turn ice in the outer system into liquid, breathing new life and vitality into those once-frozen planets and moons. Earth may be destroyed, but Ganymede or Europa might become viable homes for new life. Maybe.

The good thing about the red giant phase is that it lasts a billion years. The bad thing is that there's only a billion years before it gets much worse.

✪

All this time, while the outer layers of the sun's atmosphere have been expanding, turning red, and feasting on planets, the inner core has been contracting, growing hotter and denser, supporting itself through the strange quantum mechanical effect of degeneracy pressure. It holds itself together against the weight of a huge, freakish star ballooning above it, but not for long.

Eventually, it snaps. The pressures are too strong, the heat is too much. In a flash lasting only a matter of minutes, less time than it will take you to read this warning, the helium core goes from inert to active, slamming against itself in a nuclear fury.

This flash is especially frenetic. Usually fusion in a star's core is amazingly self-regulating. *If* the temperatures climb too high and the fusion gets too intense, the gas in the core heats up and expands, easing the pressure. Like a release valve, the relationship between temperature and density keeps the fusion rate nice and steady.

But in this helium core that tight relationship is broken, and it's all the fault of that weird quantum mechanical degeneracy pressure (which, I swear, I'll explain later). Because the core isn't supported by its own release of energy, the nuclear fusion of helium into carbon simply happens. That fusion releases energy (as is usual) but the core doesn't expand in response, it just gets hotter (as is unusual). That increased temperature (about 100 million kelvin, for those of you keeping score) makes the fusion go nuts, and it all spirals out of control, fusing entire Earths' worth of helium every single second.

In short, brutal order, the core of the star explodes and then quickly collapses, the onslaught over before it even began. A stellar Chernobyl.[6]

But to us outside the star, we amazingly hardly even notice. In those moments the core shines brighter than the combined might of all the stars in the Milky Way galaxy. But it's all for naught—at least at first. All that tremendous amount of released energy goes into expanding (finally!) and vaporizing the core.

That may not sound like much, but have you ever tried to vaporize something weighing a hundred Jupiters? I didn't think so. The core of the star acts like a massive shock absorber, taking the brunt of the helium flash (essentially the largest nuclear bomb that our solar system will ever experience) while sparing the rest of the star's atmosphere.

But in that moment, the flash of furious helium fusion, the heart of a star like our sun has its final heart attack. It will go on for a few million years, but this is the instant that you can point to as the real beginning of the end. With its heart exploded, the star is dead; the rest of the body just hasn't realized it yet.

With the core vaporized and destroyed, fusion shut off in a blink, the remaining material there recollapses. The rest of the sun, at this time a bloated and seeping red giant, follows suit. In as little as ten thousand years, the star shrinks in on itself until the pressures and temperatures in the core reach critical thresholds again to reignite hydrogen fusion in a shell, and below that helium fusion in the core.

For life in the system, everything appears normal. Or at least, a relative approximation of what was once normal. The postflash sun will be about ten times wider, forty times brighter, and burning a full orange-yellow. But, fusion reestablished, there are no extra flares, no special outbursts.

Oh, yeah, we're missing a few planets. But other than that, it's like everything used to be, back when we were young.

The new phase in the sun's life, with helium burning in the core and hydrogen burning in a shell around it, is quite peaceful. It's a shame that it only lasts for a hundred million years or so. It could've been something wonderful.

It can't last long because the sun is burning helium, not hydrogen, in the core. The fusion of helium into carbon and oxygen doesn't release much energy, so it has to burn at a faster rate to keep up with demand—the demand needed to balance the crushing gravity of the rest of the sun.

Before you know it, there's not enough free helium sitting in the core to get fusion going. It's the same story as before, but instead of hydrogen fusing to leave behind a hunk of useless helium, its helium fusion leaves behind a hunk of useless carbon and oxygen. Whimsically and perhaps running out of adjectives, astronomers call this a "subgiant" star.

And then comes the giant. Again. With fusion sputtering out in the core, some helium burning continues in a shell around it. Around *that*, any remaining hydrogen can still get its freaky fusion on. This burning shell inflates the outer atmosphere. Again. The atmosphere enlarges and cools, turning red. Again. But instead of taking a lazy billion years to get there, the beast returns ten times faster. All too soon, the sun will die its second death.

A new red giant, born from the briefly renewed sun, back for revenge.[7]

This time, it means business. The complex inner core is now incredibly unstable. The center of the star is mostly carbon and oxygen—hot, but not fusing. There simply isn't enough weight in a sunlike star to force carbon and oxygen close enough. But occasionally the fusing hydrogen layer above will drop a massive load of helium down. If it's enough, the helium will fuse and flash again, but only briefly.

The helium fusion happens faster and more furiously than hydrogen fusion, so the onion layers of inert carbon core surrounded by helium fusion surrounded by hydrogen fusion are fantastically unstable. The helium shells expand, bursting through the hydrogen layer. But higher up in the sun's atmosphere there isn't enough helium to keep the fusion going, and so it fizzles out. The core collapses (again).

But then the hydrogen fusion builds up another critical mass of helium, leading to a new round of activity.

With each new cycle, the sun will shrink, having a new source of fusion at the core, then expand again as fuel runs out once again. Pulsing like a giant heart at the center of the solar system, switching from normal white star to furious red giant over and over again, repeating itself every hundred thousand years.

This new red giant phase is defined by its extremes: the sun will swell to the radius of the Jupiter's orbit before shrinking back down to something almost as small as it once was in the placid hydrogen-burning main sequence days. Days now a distant memory, lost in the fog of violence.

Rapid shrinking, rapid expansion. Sputtering and gasping fusion in the core. Spasms on the surface. Sunquakes. Material will be ejected in massive solar storms. Coronal mass ejections may be bad during the years of the normal sun, but in this phase they're like hailstorms compared to a light drizzle. Even the outer planets will suffer extreme magnetic and cosmic ray storms. This is an especially difficult time for life, and for explorers.

Stars in this phase are especially dangerous. Called Mira variables, after the star Mira, the first of their kind to be found, they can change their brightness by a factor of ten in as many days.[8] Within a couple weeks, a Mira variable and grow ten times brighter and ten times dimmer. Swelling, expanding, contracting, shrinking. Repeated ad nauseum. Pulsing and churning, sending their systems into total chaos.

The name Mira itself means "amazing one," when astronomers hundreds of years ago watched over the course of only a single year one of the brightest stars in the sky fade to invisibility. Mira still exists—you can see it with powerful telescopes and visit it if you're foolish enough—but to those ancient astronomers, a familiar light in the night sky simply winked out of existence.

Amazing, indeed.

Mira isn't alone. R Hydrae, S Carinae, U Orionis. Hundreds more are known and mapped; millions more unknown flare and gasp in the vast expanse of night that is our galaxy.

And all the while, while the star loses its very sense of self, the core of carbon and oxygen builds and grows. Silent, waiting. The final remnant of the sun is coming into its own. For now, it is hidden behind the fiery unstable veil of the turbulent atmosphere.

For now.

✪

Before the sun dies its final death, it heaves one last great bucket of vomit into the system. Gross, I know, but pretty accurate.

After several phases of these pulses, the sun has already lost quite a bit of its mass, especially during the extra-energetic ejections. But there's still plenty more to go. Eventually, one of the flares goes too far: too much material falls in at once, ignites, burns too quickly, and flashes, sending the rest of the sun's atmosphere into space. Like a nuclear-powered bouncy ball.

At first the ejected plasma just hangs out nearby the sun, undecided between reuniting with the core or going for broke. Two things make the decision for it. An extreme form of stellar winds—the usually steady stream of tiny particles spit out by the sun—pushes that layer of material out and out and out.

The other is the intense radiation beginning to make its way out of the core and through the swollen and gangrenous outer layers of atmosphere. As the outermost layers escape into space, beginning to separate from the core, then become cool enough to form atoms, the first time they've experienced that state of matter in billions of years. In that atomic state, the gas becomes thick to the radiation pouring out within it. It absorbs that radiation, heating up and expanding outward. In the process, it disassociates again into a plasma, temporarily pausing the expansion.

But gasp by ragged gasp, the sun turns itself inside out, pushing almost half of its raw material out, beyond the former inner planets, past what's left of the asteroid belt, and all the way out into the furthest reaches of the system.[9]

I should note that as the sun goes through this roller-coaster diet phase, the planets don't just sit there ignoring it. Every time the sun pulses or swells up to a red giant, or expels another plume of hydrogen, their orbits get a little tweaked. A few tweaks here and a few tweaks there, and by the end of the process nothing's like it used to be. Sure, some planets might survive—and we see evidence of those planets in nearby systems. But the ones that hang on are just the dense, rocky cores of what once were majestic gas and ice giants, reduced to ruin and misery as they watch—helplessly trapped by bonds of gravity—to watch their mother star die.

And yet, in all this chaos and gore, the dying sun puts on one final show. One last performance before permanent retirement. One final goodbye to the galaxy, reminding everyone how beautiful a star can really be.

In its final stages of life, after billions of years of complacent hydrogen fusion, a billion years of runaway buildup, a flash of furious intensity in the form of a helium flare, a hundred million years of bedridden despair, and the final plunges of violent expulsion of its own body, the sun becomes a *planetary nebula*: the guts of a sunlike star speckled across its former system. Up to half the mass of the original star. *Half.* The sun is, by far, the most massive object in the solar system, and the final phases of its life—the last big flashes—had enough energy to send half of it into deep space, never to return home.

OK, I admit, the name is a little weird. They don't really have anything to do with planets. They were first discovered by Earth astronomers once their telescopes became big enough to see them, and they kinda sorta looked like roundish fuzzy patches. At least, they looked like that in the miserable telescopes they had at the time. The first astronomers to find them were smart enough to figure that they weren't planets, but since they were somewhat planetlike they figured they might be capturing the formation of planets around young stars. And *nebula* is Latin for "cloud," so it seemed to fit.

Now, much smarter and wiser than our ancestors, we know that they had it completely backward: these aren't the beginnings of a star's life, but the end. But we're stuck with the name, because of history.

Anyway, gas floating around the solar system—even gas that was once a part of the sun—isn't enough to make a spectacle. No, to make a really big show you can't just have camera and action. You need *lights*.

The lights come from the core. The remnants, the leftovers, the forgotten husk of carbon and oxygen. After fitfully ejecting the rest of the sun's material into space, it now sits exposed, naked and alone. And it's *hot*. Not quite hot enough to start a new batch of fusion (we'll talk about that particular nightmare later), but really dang hot, and bright—four thousand times brighter than the sun is today, containing nearly half of the sun's original mass, converted into almost pure carbon and oxygen in a volume no bigger than the (now former) planet Earth.

This is the beginning of the *white dwarf*, the leftover bones of the sun that will last for billions, even trillions of years. I'll get to them in more detail too, should you ever be foolish enough to encounter one, but for now all we care about is their heat. For only a few tens of thousands of years, the leftover core is hot enough to spit out deadly X-rays, soaking the spewed-out guts in all their radiant glory.

When you slam an atom or molecule with high-energy radiation, a few things can happen. It can be kicked around. It can lose a few electrons. Or it can glow. The atom can absorb the light, soaking up the energy, and spit it back out at a different frequency. Different frequency means different color.

This is how neon signs work: electrocute a tube of gas, filling it with energy, and make the molecules glow. Depending on what they're made of and the amount of energy you pump in, different elements will have different signature colors. Hydrogen is red, helium is yellow, mercury is blue, and the classic neon is orange.

The same physics that lights up Vegas and Times Square lights up the leftover dead solar system. The sun's material, ejected in multiple episodes, tangled up with itself, twisted around fleeting magnetic fields, gets one brief moment of glory.

Lit up from one end of the former solar system to the other. Intricate lattices. Streamers of gas millions of miles long. Puffy egg-shaped prominences. Hourglass figures. Layers and layers of cotton candy. Dazzling colors as the trace gases absorb and reemit the radiation in the own unique way.

A jewel against the night. A work of art, one last expression. Ten thousand years, maybe a little more, maybe a little less. But that's all it gets. All too soon, the white dwarf cools, no longer emitting the needed high-energy radiation, and the nebula fades back into darkness, its moment of glory gone forever.

✪

Their beauty is what draws explorers into their embrace. Their deadly, poisonous embrace. The radiation needed to light up a massive nebula is enough to kill an unprotected traveler. X-rays. Gamma rays. The hard stuff. This radiation won't just give you a little tan, it will peel off layers of skin, it will roast you alive.

The creation of the nebula, too, as fascinating as it might be to watch, causes extra harm. The stellar winds that drive the inflation of the gas, that shape it into beautiful, complex forms, are cosmic rays on steroids. The tiny little particles pack enough punch to shove half the sun to the very edges of the solar system. What do you think it would do to you? Or a ship? It will happily push it around like the tin can it is.

In its death throes, just before and during the creation of the planetary nebula, a star is at its most schizophrenic, switching chaotically from stability to catastrophe in a matter of days, blasting the system with renewed bouts of cosmic rays and radiation. Catching unsuspecting explorers and probes unawares. The stellar winds can go from a breezy 5 miles per second to a hurricane of 1000 miles per second in a matter of days.

The nebula itself is rather safe: it's just a cloud of hydrogen and a few other bits. Once ejected, it slows and cools quickly to safe and reasonably tolerant levels. Nothing too bad, I would say.

But it's the leftover star. The brief moment between its unveiling and its cooling, when it's lighting up the nebula like a neon sign, is when things get scary. When you've got to watch your back and watch the clock. When one moment you're enjoying the beautiful, transfixing view inside the nebula, and the next you're ashed in the next blast.

Planetary nebulae are indeed lovely. Part of their loveliness is their transitory nature: they'll only last for ten thousand years, give or take. But since

A lonely mountain on the surface of the Earth, photographed in artistic black-and-white? No, a close-up view of comet 67P/Churyumov–Gerasimenko (say *that* three times fast) as imaged by the plucky Rosetta spacecraft. This little comet is just a few miles across, and spends most of its time wandering between Mars and Jupiter. If you ever get this personal with a comet, let's just say that you're in for an adventure. Yeah, adventure. *Courtesy of ESA/Rosetta/MPS for OSIRIS Team MPS/UPD/LAM/IAA/SSO/INTA/UPM/DASP/IDA.*

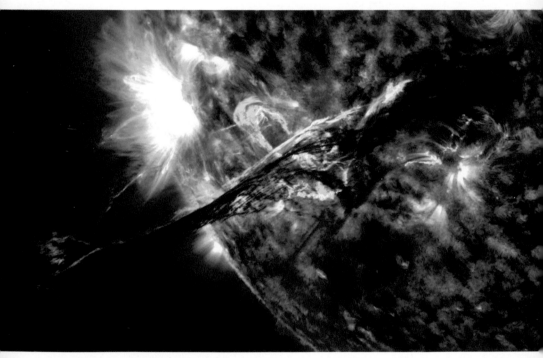

The normally even-tempered sun on a very, very bad day. This particular lovely prominence—an arc of plasma launched by hyperactive magnetic fields—could engulf the entire planet Earth several times over. And you thought your toddler's temper tantrums were bad. *Courtesy of NASA.*

The beautiful, effervescent, delicate Orion Nebula. Birthplace of a batch of new stars. It's also a hotbed of high-energy radiation, shock waves, and exploding stars. Please, observe from afar. Trust me, it's not as pretty on the inside. *Courtesy of NASA, ESA, M. Robberto (Space Telescope Science Institute/ESA) and the Hubble Space Telescope Orion Treasury Project Team.*

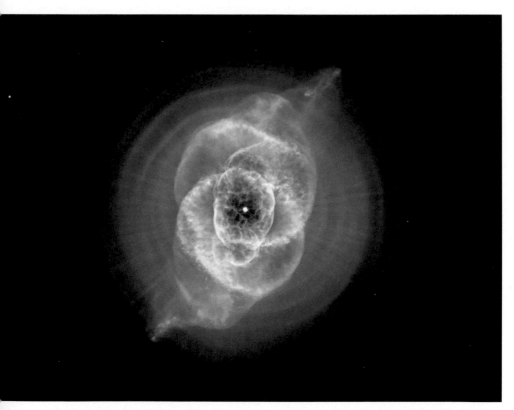

And here, on the opposite end of the nebula spectrum, not the birthplace of stars but a gravestone: a planetary nebula. This one, unromantically designated NGC 6543 (but the truly poetic explorers will always call it the Cat's Eye Nebula), took about 1,500 miserable years to form as its central star turned itself inside out. Yes, it's pretty, and yes, it's gross. *Courtesy of NASA, ESA, HEIC, and The Hubble Heritage Team (STScI/AURA).*

Another tomb: the Crab Nebula. I suppose it kinda-sorta looks like a crab, if you squint your eyes and look through a 19th-century telescope. Anyway, it's the leftover debris from a titanic supernova explosion, one bright enough to be seen in the daytime in the year 1054 CE. At the center sits a pulsar, the last remnant of a once-great star. The rest? A toxic wasteland of radiation and failed dreams. *Courtesy of NASA, ESA, J. Hester and A. Loll (Arizona State University).*

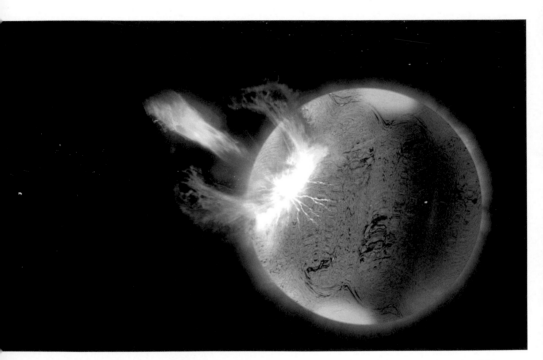

Ah, magnetars, the fast-spinning zombie stars of the cosmos, looking alive when they should be dead. A face not even their stellar mothers can love. This is very obviously an artist's rendition of a flare produced when their crusts crack like the top of a good crème brûlée, but without all the sugar. Why is it obviously an artist's rendition? Because no probe or explorer could survive such a blast. *Courtesy of NASA's Goddard Space Flight Center/S. Wiessinger.*

This poor star! It orbits, helplessly, as its now-dead sibling feasts on its atmospheric flesh. The black hole eats and eats, never sated, drawing in ton after ton of gas into its ever-gaping black maw. We can see the screams of agony and cries for help from thousands of light-years away, as the gas reaches blisteringly hot temperatures before falling into the eternal abyss, never to be seen in this universe again. *Courtesy of NASA/CXC/M.Weiss.*

The accretion disk. The jet. The event horizon itself. These are the monster black holes at the centers of galaxies; the most fearsome engines in the known universe. Quasars, blazars, call them what you will; a single black hole like this is capable of reshaping entire galaxies. They are, to put it simply, scary. If you want to visit one, be my guest, but I'll stay safely over here. *Way* over here. *Courtesy of NASA/JPL/Caltech.*

almost every star ends up as one, they're as common across the galaxy as ornaments on a Christmas tree.

But just like ornaments, they can and should be enjoyed from a distance.[10] They may be pretty, but they're nasty, too.

Besides, by the time you get to one, it will probably already be gone.

White Dwarves and Novae

Several times I've brought up this weird phrase: *degeneracy pressure*. Technically speaking—and you'll have to allow me to be technical here—it's a pressure exerted by degenerate states of matter.

Wow, don't you feel enlightened now.

Fine, I'll dig a little deeper, just this once. And we're going to dig really deep. I am, after all, trying to educate you not only on the *whats* of the dangers of interstellar adventure, but also the *hows*. But if you're content to simply say to yourself that "degeneracy pressure is a way that a dead star can hold itself up against the collapse of gravity" and move on with your life, I'm not going to get in your way. Just skip these next few paragraphs and mosey on down to the next section, where you'll find all the usual dangerous physics you were hoping for.

But for the truly brave—the ones who are willing to go to the unknown regions of your mind and open doors of knowledge that most are too scared to even approach—read on. Your courage will be rewarded.

Degeneracy pressure. It's here and it's important.[1]

First, we need to acknowledge the fact that our universe is split into two distinct tribes of particles. Particles like electrons, quarks, and neutrinos are called *fermions* or *spin-½ particles*. These are the familiar building blocks of the world around us: if you zoomed way, way, way into your own cells, you would find atoms, which have a bunch of electrons orbiting a nucleus. That nucleus would be made of protons and neutrons, and those protons and neutrons are little balls of quarks.

What are the quarks made of? Well, a) it doesn't look like they're made of anything, and b) it's not our problem.

Also, neutrinos never do anything important, except when they do, but we'll get to that when we look at how the biggest stars die.

The other kind of particles are called *bosons* or *spin-1* particles. These will be the photons, gluons, W bosons, and so-ons. You might or might not recognize these as the carriers of the fundamental forces, the most famous of which of course being the photon, responsible for everything from the warmth on your skin during a beautiful summer day to the acute radiation poisoning from a hard dose of gamma rays.

The fermions are named after Enrico Fermi, and the bosons are named after Satyendra Nath Bose. Great.[2]

And the spin-½ versus spin-1 business? No, I will not explain that, thank you for not asking. If you're *really* burning with curiosity, well at least now you know what to type into your favorite search engine.

Why is the universe this way? Why do we have this split between fermions and bosons? I don't know, that's not my job. My job is to explain to you why you should care.

Say you're holding a bag of fermions. Let's go with electrons, they seem common enough that this shouldn't be too hard. It doesn't have to be electrons, it could be protons or quarks or whatever, as long as it's not photons or one of the other bosons. They play by different rules, and what we're talking about only happens with fermions.

That bag is a like a little hotel: electrons check in and get put in a room. Physicists have a very specific word to describe the rooms in this hotel: they call them *states*. A state is a complete description of the status of each electron: its energy level, its spin, and whatever else you might need to care about when you're sorting electrons into rooms.

So it just so happens, for various and somewhat mysterious rules of the universe, that electrons and their fermionic brethren hate sharing rooms with each other. They like their own bed with their own shower and I've heard that the top quark snores like a chainsaw. One electron per room, max. One fermion per state, max.

(The bosons, by total contrast, love to pile in as many of themselves as they can into a room. The more the merrier. They love to party. I know, they're weird, and we won't talk about them further.)

So as the electrons check into their little hotel, they start with the first room on the ground floor, then the next room, and so on, until all the rooms on the ground floor are filled. Then they go to the second floor, and up and up and up until you've put every electron inside a room.

The size of the hotel and the number of rooms in each depends on the exact physical situation. For example, for electrons in an atom, there are two rooms available on the ground floor and eight in the next, and more as you go up. This means that only two electrons can live closest to the atomic nucleus, with eight electrons living in a higher-energy state above them. This restriction on the number of rooms in the atomic hotel sets the rules on how atoms can share electrons and combine with other atoms to form molecules. You know, chemistry.

This also explains why stuff in space takes up space. Have you ever thought about silly little questions like, "Why do I have volume?" No, it's not the potato chips, although that certainly doesn't help.

Without these rules about how electrons hate to share rooms, the negatively charged electrons in an atom would simply immediately crush into the positively charged nucleus, because of their mutual electric attraction. Instead, the electrons are forced to stay inside their hotel rooms, and those rooms are far away from the nucleus. In other words, an electron simply can't get closer to a nucleus, because it can only go in the first room on the first floor of the hotel—that's the bare minimum, lowest-energy state allowed. And so, atoms have volume, and everything that's made of atoms (for example, you) also has volume.[3]

For the electrons you're putting in a bag, you'll find that the electrons don't all settle into the bottom—they take up space. Now there are a few reasons why the electrons in your bag, as opposed to the electrons in an atom, take up space. One reason is that all electrons are negatively charged (it's kind of a part of the definition of *being an electron*). So they naturally can't stand each other anyway, by nature of their electric repulsion. But this only becomes an issue when the electrons are really, really jammed in close together—otherwise their electric fields are just too wimpy to really be much of a player.

Another reason why the electrons in your bag take up space is their temperature. Cold things don't move around a lot, and hot things do. So you can imagine heating up your bag of electrons, and they start whizzing around all over the place, bouncing off the inside of the bag and careening off each other. With all that extra energy, they're bound to take up a great volume.

But here's the thing about electrons and their extreme distaste in sharing a room: it applies even at a temperature of absolute zero. Now I know just as well as you that absolute zero is an impossible temperature to reach, but work with me here for the purposes of this illustration.

If you have a hot bag of electrons, the rooms (or *states* if you're getting tired of analogies) at the ground floor of the hotel won't even be filled—all the electrons will have nice, high energies and enjoy floating around the penthouse suites rather than the grungy rooms with nothing better out the windows than the parking lot. The same quantum rules apply, as always—the electrons still won't share a room—but they'll just hang out at the upper levels.

As you cool your bag down, the electrons lose their ability to reach the highest floors, and start to populate lower and lower levels of the hotel. Some may even—gasp—be forced to take a spot on the ground floor. Those plebes.

Anyway, at true, real, actual absolutely absolute zero, the electrons don't have any extra energy at all. They will fill up the rooms in the way that we started this whole tortured analogy with—with the first room on the ground floor, working up from there. The last electron to go in the bag will simply sit on top of all the other electrons.

And here's where the magic of degeneracy pressure kicks in. Imagine trying to squeeze and compress a giant ball of electrons. If the electrons have any temperature at all, they'll resist you by flinging themselves against your skin as hard as they can, offering a source of pressure to resist that collapse. This is called thermal pressure, because it has to do with temperature.

Squeeze them more, and their natural electric repulsion will also try to fight you. This is called electrostatic pressure and is, like I said earlier, not that much of a player in the kinds of systems we're about to explore, but I wanted to get it out there for the sake of completeness.

The electrons have been fighting you, resisting you in your attempts to squeeze them further. But you're clever: you dropped their temperature to absolutely zero (I know, technically impossible, but we're in the magical Land of Fantasy for the Purposes of Illustration here). The electrons put up less and less resistance, less pressure, as you continue to squeeze.

But then the rooms fill, one by one, from the ground floor on up. At zero temperature, you have a bunch of electrons occupying as many hotel rooms as they can—no more, no less.

You push some more. What happens?

Nothing.

Nothing happens. You press on the electrons, and there is *literally nowhere else to go.* They don't just hate sharing rooms (or states), it's almost physically impossible for them to do so. You press onto them, wanting them to compress more tightly, and they just look at you, shrugging, wondering to themselves when you're going to realize that this is a pretty useless exercise.

The electrons are able to offer a resistance to your compression, through nothing more than a quantum mechanical fact of nature. It's a pressure. A degeneracy pressure.

Now the name *degeneracy* is a little bit odd—I haven't even come close to defining that word or explaining why it would appear in this context. So here we are. In the weird, wonderful world of quantum mechanics, sometimes a single energy level can have multiple states. For example, two electrons in an atom can have the same energy—in this case, the same distance from the nucleus—but in one of two distinct cases: one with a spin pointing up and one with a spin pointing down.

What is spin? Don't worry about it. If I dig into *that* one we'll be here until the end of the book. The point is that when a single energy level (like a floor on a hotel) has more than one available state (like different rooms on the same floor), then that energy level is said to be *degenerate*, and the electrons in those states are also said to be *degenerate*.

Degenerate states. Degenerate matter. Degeneracy pressure.

Just in case you didn't quite grasp how degeneracy pressure works, here's another way to look at it. It uses a completely different language and line of argument than the metaphor I just used above, but mathematics is mathematics and oftentimes in physics we find ourselves with the freedom to describe the same phenomenon from different angles.

And so, a different angle.

Have you ever heard of the *Heisenberg uncertainty principle* (sometimes tragically abbreviated as HUP[4])? It's another one of these weird rules of the quantum world that make no sense to us macroscopic beings because we're not quantum things and we operate by a different set of rules that got hardwired into our brains as The Way Things Are. The HUP (yup, I did it) is a fundamental relationship between the momentum and position of a subatomic particle. In particular, it tells you what you *can't* know about that particle: the more you know about its position, the less you know about its momentum, and vice versa.

In other words, there's a fundamental quantum limit to knowledge. Don't like it? Too bad—this is our universe and we have to play by her rules.

When you squish down electrons to be all tight and compact against each other, your knowledge of their position grows. Instead of being vaguely "well they're around here somewhere, I guess" with a general wave of your hand, they are precisely "right there, where I put them" with a firm a decisive point of your finger.

But the HUP tells us that the more you know about their positions, the less you know about the momenta. The more you put them in a corner, the more fiercely they vibrate and wiggle around. Like trying to put an angry bee inside a little box, and the smaller you make the box the angrier the bee gets.

That vibrating, that buzzing, that uncertainty in momentum manifests as something fighting against you, which you feel and recognize as a pressure. Try to compress your degenerate ball of electrons down even tighter, and it will fight you even more. Again, from a weird facet of the quantum universe.

Why do we care about degeneracy pressure so much? Because it's these strange rules of quantum mechanics that govern how white dwarves, neutron stars, and a few other weirdos in our universe are able to hold themselves up.

In the normal life of a star, the energy released from nuclear fusion is enough to keep things hot and provide thermal support against the never-ending crush of gravity. We saw how things start to go haywire inside a star once a nugget of degenerate matter appears and throws off this detailed balance. The main issue with that nugget of degenerate matter—and one of the key features of all degenerate matter, everywhere—is that it behaves exactly the opposite of the way it should when you add more mass to it, and when it gets hot.

Usually when you add stuff to a thing it gets bigger. Add a pebble onto a mountain, and the mountain is now bigger, by exactly the size of the pebble. Add a piece of cheese to your stomach, and you are now larger by the size of a piece of cheese. Kind of obvious. But for stars made out of degenerate matter, if you add more stuff, they actually get *smaller*.

This works because nothing is quite at absolute zero, so not all of the lowest floors at the quantum hotel are filled. When you pile on more mass, there's more gravitational pull, because of the extra stuff, and because of all that extra weight some more electrons get shoved down into the lower floors. The degeneracy pressure is still able to hold up the star, but its volume does contract: one side of the teeter-totter tilts down, and you end up with smaller star. Neat.[5]

The other weird thing about degenerate stars is that they don't get bigger when they get hotter. I know, how annoying. Usually, when you heat up a gas, all the particles zoom around more freely and vicariously, taking up greater volume. But gravity is still gravity in a degenerate star, holding everything together, so when you heat up a giant ball of degenerate matter you end up with nothing more than a hotter giant ball of degenerate matter.

All the rules of thermodynamics that we're so used to in our lives, and that govern the lives of stars from before the moment of their births through millions and billions of years, just give up and hand over the keys to quantum mechanics, which is generally known for its disorderly and reckless conduct.

★

OK, enough fooling around. If you skipped the last section on degeneracy pressure: hello, welcome back. Good to see you. If you stuck around and read the whole thing: congratulations, you are now a degenerate. That's supposed to be a compliment.

Let's get down to business, and in particular, the business of white dwarves. "Dwarves" because they're smaller than the average star, and "white" because they're white. Done.[6]

These are the first kind of strange, exotic, quantum degenerate objects that you're likely to meet in your travels around the galaxy. I've already told you how they're born, and like most things in the universe, their birth is the consequence of the death of something else. A regular star, much like Earth's own sun, will eventually exhaust its supply of juicy hydrogen and helium, forming an inert lump of carbon and oxygen deep in its gullet. Unable to digest it, the star explodes (well, not so much *explode* as *violently spasm*), scattering its guts deep into space, leaving behind the white hot core.

White dwarves are born from stars like our sun, while black holes are born from much heavier things (we'll get to that in detail soon, don't worry my eager little explorer), and since stars like our sun are much more common in the universe than much heavier things, you're bound to see and encounter them much more often than their blacker brethren. Indeed, the nearest one to Earth isn't even ten light-years away, a companion to Sirius, the brightest star in our home world's sky.

Let that sink in: Despite their fearsome reputation, you'd have to trawl in the deep interstellar depths to find your first black hole, and then it would only be from (bad) luck and not from any intentional hunting. But a white dwarf, a dead core of a star held up by only exotic quantum forces? You can barely make it out of our solar system's front door before tripping over them. Rough estimates guess that just our Milky Way galaxy is home to around 10 billion of these things.

And yes, they're deadly.

Let's talk about their density. When they were first discovered by Earth astronomers in the early 20th century, they seemed downright preposterous, if not outright impossible. A single object, weighing as much or sometimes more than the mass of our sun, compressed into a tiny ball no bigger than the planet those astronomers were standing on. This rounds out to an average density about a million times greater than water. What the heck was going on? Science is the job of listening to the secrets whispered by

Mother Nature, and apparently when she finally told us about the wonders of white dwarfs, our first response was to tell her to "shut up."[7]

The gravity at the surface of a white dwarf is so intense that, were you to find yourself stranded on it, you would need a rocket capable of 2 percent the speed of light to escape.

This ridiculous, awe-inspiring density comes from their degeneracy pressure. Inside a white dwarf, all the electrons have been torn from their atoms, woken up in the middle of the night, and kicked out of their cozy atomic hotels. They wander free in the streets, looking for any shelter they can find. They may find themselves in a new environment, but the old rules still apply. They can't share a bed. Find a cardboard box beneath the underpass? Just one or two electrons can shelter there. Squatting in an abandoned house? Just a few newcomers, please.

The electrons in a white dwarf star can only compress *so much* before the rules of the game simply prevent them from squeezing further. And that's enough to support the star. Usually it's the explosive energies of nuclear fusion that resist the constant squeeze of gravity, but now it's simply good old-fashioned quantum mechanics. The best kind of mechanics.[8]

A white dwarf is a collection of loose carbon and oxygen nuclei, swimming free in an endless sea of tightly compressed electrons. A sweltering, stuffed, cheek-to-jowl quantum mosh pit. The degeneracy pressure of the electrons keeps the star supported against further compression, and it's a much more compact configuration than a normal star can ever hope to achieve.

But there is a limit.

One way to understand the maximum mass of a white dwarf is to remember the bit about the Heisenberg uncertainty principle. Oh, you skipped the last section and don't remember it? Too bad, you've got to know this stuff anyway if you're going to have half a chance at surviving out there. The HUP told us that the tighter we squeeze on the ball of degenerate matter, the faster the electrons wiggle around, leading to the familiar pressure.

But electrons can only go so fast: they're limited, like anything else, by the speed of light. Eventually, if you squeeze and squeeze on your white dwarf, the electrons would have to break the light barrier to keep resisting you. They can't, so they don't, and the white dwarf collapses under its own weight, the saving grace of quantum degeneracy pressure unable to keep its end of the bargain.

The first person to estimate the maximum mass of a white dwarf was Subrahmanyan Chandrasekhar, and the Chandrasekhar limit is dutifully named in his honor. The Chandrasekhar limit is 1.44 times the mass of the sun, according to his first calculations, and revised a bit downward with more sophisticated analyses that include effects of rotation. But that's it: if you try to build a white dwarf bigger than that, it will catastrophically collapse in on itself, with delightfully explosive (and fearsomely dangerous) results that we'll get into later.

For now, we'll focus on the white dwarves themselves.

And especially their heat.

When they're first revealed, the white dwarfs are intensely hot. It's not generating any of its own heat, but come on, it's the exposed *heart of a star*; it kinda has a lot of heat getting started. Before the star went through its final death-throes, turning itself inside out in a display of pure gruesome beauty, its heart had long ago stopped beating. But the constant onslaught of gravity and twin burning shells of helium and hydrogen fusion dumped kelvin after terrible kelvin into its core, enabling it to reach a breathtaking ten million degrees.

What's most amazing about this is that the white dwarf will be able to maintain that temperature for billions, and even trillions, of years without fusing a single atom of hydrogen or helium. With no source of heat, eventually the white dwarf will indeed cool down, but the only way it can do so is through the radiation of light into the vacuum of space. The body of the white dwarf quickly accumulates a faint, hazy atmosphere of nondegenerate (in other words, boringly normal and not quantum at all) hydrogen and helium. It's this atmosphere that is actually exposed to space and has to do the job of cooling off the white dwarf, but at the moment of their birth the surface has the same temperature as the core (read: too hot). It eventually cools to just a mild few thousand kelvin.[9]

(And you know we're doing some serious interstellar wandering when a few thousand kelvin is considered "mild.")

So despite their sweltering interiors, they have relatively cool surfaces, and because of their tiny stature it takes an achingly long time to fully cool off. As they do, they gently decrease in brightness and shift in color from intense blue-white down into cool and calm reds, and eventually infrareds.

How long does this take? To give you some sense of the timescales we're talking about here, our universe has an ancient and wise age of 13.8 billion years. The first stars, and hence the first white dwarves, appeared within the first few hundred million years into our existence. The coldest white dwarf known still has a temperature just under 4,000 kelvin—our universe simply hasn't been around for long enough to achieve anything colder.

So the general rule of thumb for our present universe and for roughly the next trillion years: if you see a white dwarf, *it will be hot.*

It's hard to describe what it would be like to live near one. But I'll try. Imagine the sun shrunk down to the size of the Earth, nothing more than a comparatively tiny point, but shining a hundred times brighter than it does today.

"Intense" would be a good word for them. Yeah, *intense.* Say it like you mean it.

The first ten thousand years are the worst. That's when their surfaces are hot enough to emit the X-rays needed to fluoresce the planetary nebulae around them. After that, they cool to the surface temperature of a typical run-of-the-mill star and just hang out there for eons. This may make them seem relatively harmless—after all, black holes are monsters for essentially forever, so this seems rather brief as hazards go—but remember that new white dwarfs are being exposed all the time throughout the galaxy. You may be heading to what you hope to be a curious late-stage main sequence star, and by the time you get there you get a face full of hard X-rays.

As the white dwarf cools, the carbon and oxygen atoms inside it will naturally arrange themselves into a crystal lattice, because that's what carbon and oxygen atoms like to do when they're nice and cold. You may recognize "carbon organized as a crystal lattice" by the more familiar term of *diamond.* A diamond the size of a rocky planet. Pro-tip for the lads: this would make an *excellent* Valentine's Day present. Although, you would have to wait a few years for the star to become cool enough to handle.[10]

And by "a few" I mean at least 10^{15}.

In the meantime—and there's a whole lot of meantime—we have to deal with them. By themselves they're not entirely dangerous: just a white-hot source of deadly X-rays in their local systems. But we've already dealt with intense sources of X-rays, so you should be pretty well equipped to handle them. Just stay out of their local systems, and you're good.

But while white dwarves may seem dead—the leftover husks of once-thriving stars—they are far from their final rest. Underneath their calm surfaces, they pulse and hum and vibrate. They're *alive*. Or at least, undead. Strange forces in their interiors flex and pull and push and prod. They're not completely solid, not yet, and waves can slosh back and forth across their surface and deep into their cores. Sometimes these waves peak and crest, dredging up heat from their innards and exposing it on their faces. In a matter of minutes or even seconds, a white dwarf can brighten. Not by much, maybe by a quarter or a third, but it might be enough to launch a new round of X-rays, stronger than you predicted. A star that may seem dead and quiet may suddenly flare, a last grasp for long-forgotten vitality.

✪

White dwarves in isolation don't pose that exotic of a threat, despite the strange quantum forces supporting them. But these creatures become especially dangerous when they have a companion, a partner in crime.

Take two stars. One is slightly fatter than the other. Don't judge. The bigger star burns hydrogen at a faster rate, going through its life cycle: increased brightness, red giant, ejected material, white dwarf remnant. You know the drill. It's settled into a comfortable retirement, slowly growing dimmer as the eons pass by. It's had its moment to shine against the night, time to pass the torch to a new generation.

But its companion is just getting warmed up. It has the same life cycle, but being less massive, it goes through everything slower. It's still steadily burning hydrogen in its core as it watches its lifelong companion go red and leave behind the white dwarf. After billions of years of orbiting the silent white dwarf remnant, the companion star goes through its Big Change too, flaring up into a red giant.

We've all seen those buddy cop movies. One is a grizzled veteran on the verge of retirement. He carries around pictures of his dead wife and/or grandkids. The other is a hotshot new recruit, eager to hit the streets but unwise in the ways of the world. Hilarity and/or hijinks ensue.

The white dwarf of the pair is stable: it's a solid hunk of carbon and oxygen, held up by that lovely degeneracy pressure. It can—and will, given the choice—hang out in that state for trillions of years, slowing cooling off and sliding into utter blandness.

The less-massive star, the young hotshot, has inflated to be many times its original size. Material from that new red giant is oozing across the stellar system, with some of it lured by the gravity of the white dwarf and finding its way onto the surface of the silent companion (this is the exact same scenario that we encountered earlier when it came to massive stars orbiting a black hole, so don't freak out if you're getting a sense of déjà vu here). As with most things in the universe that move, it goes in a circle, swirling around the white dwarf in a bright, hot accretion disk before ramming into the surface itself.

That disk itself, like disks across the cosmos, is a nasty place. Get caught up in that blender and you're bound to get vaporized.

But as for the white dwarf, at first it's no big deal. What's the problem with a little hydrogen blanket on a cooling mass of carbon and oxygen anyway? There's no problem at all, until that blanket gets a little too suffocating.

Bit by atomic bit, eventually enough material can pile up around the white dwarf to crushing pressures, squeezed from above by the ever-present force of gravity, and heated from below by the smothered and annoyed white dwarf. This thick atmosphere can build over the course of days, years, decades, even centuries as the red giant pours its own material onto its long-dead companion.

And then too much becomes too much. The temperatures and pressures in the newly constructed hydrogen atmosphere of the white dwarf reach those certain special critical thresholds, and the dance of protons begins as they find each other, overcome their natural repulsion, and smash together, forming helium.

A flash of fusion, not seen in the white dwarf since the glory days of main sequence burnings. But usually fusion happens only on the *inside* of stars, because it's really hard to do and stars are the only thing big enough (by definition) to reach the incredible levels needed to do the trick. But white dwarves are so dense (remember, you're packing a sun's worth of stuff into a ball only the size of the Earth) that their surfaces have enough gravitational pull that nuclear fusion on their surfaces is pretty much inevitable.

The flash on the surface of a white dwarf is called *runaway fusion reaction* and that's exactly what you should do.[11]

The initial flash, sparked in some random location by pure unlucky chance, spreads like a flame across the surface of the white dwarf, setting fire to the hydrogen, fusing it into helium, and releasing a burst of energy and radiation. This burst of energy triggers a new round of fusion in its wake, in short order completely engulfing the surface of the white dwarf in a blaze of fusion glory.

A *nova*.

So named by Earth astronomer Tycho Brahe, who recorded in detail the appearance—and eventual disappearance—of a *stella nova* (a "new star") in his sky.[12] So much detail that he wrote a whole book about it, just to show off how great his observing skills were. He never knew what caused this nova, and for centuries astronomers applied the name to any old random thing that flared up in the night sky.

Today, with the benefit of a few generations of insight, we now know that Brahe didn't see what we call a nova, but a *super*nova, which we'll talk about later. I know it's a bit confusing, but in the early 1900s astronomers decided they needed to sort through all this complicated cloud of jargon words, and ended up just making an even bigger mess of things. Nowadays we just leave well enough alone.

The proper astronomy conference term for a nova is *cataclysmic variable star*. That right there should trigger your little mental alarm bells: cataclysmic. Yikes. Not a word you want to mess around with. But most people just call them novae (or novas) anyway because "cataclysmic variable star" has too many syllables in it.

But even without the "super," a nova can be pretty nasty, too. These are highly unpredictable: the chance of a nova occurring is bound to happen with this binary star setup, and binary stars are an incredibly common situation in the galaxy, but you don't know when and you don't know how. A pretty-looking binary system with a white dwarf paired to a red giant can blast you while you're still in the middle of your last blink.

And when it goes, it's like a multi-multi-multiton thermonuclear bomb going off. Imagine turning the entire Earth's atmosphere into a weapon. Now you get the idea. These novae throw off so much heat and light that they can be seen from thousands of light-years away. Every few years, Earth-based astronomers are treated to a nova so bright it can be seen with the naked eye. The heat and intensity from the blast will last for around a month or two, as the exploded remnants of freshly fused helium heave throughout the system at thousands of kilometers per second, irradiating the local system before it cools down again.

And a white dwarf doesn't even have to go full nova to make your life miserable. All that material falling onto the white dwarf forms an accretion disk. The accretion disk maintains a nice steady drip line, feeding a thin stream of gas continually onto the white dwarf. But every once in a while, the accretion disk can get a bad idea and decide to collapse altogether onto the white dwarf. That's a lot of gravitational energy released at once. It may not go nuclear but it sure does make a nice hot flash for your enjoyment and/or horror.

Like I said, we're not exactly sure when a particular white dwarf might go nova. It can even do it more than once, if it feels like it. A typical white dwarf locked in a gravitational embrace with a red giant will have a major flare-up every few thousand years. But since very little of the white dwarf itself is damaged in the event—all the action takes place in the hydrogen blanket it built up from its hapless companion—some have been known to have violent fits every few decades. In a galaxy like the Milky Way, there are a few dozen novae every single year. Thankfully it takes a very special combination of factors to ignite them, otherwise the galaxy would be a very dangerous place indeed.

Well, more dangerous than it already is.

✪

While we're on the subject of things orbiting each other leading to mayhem, let's talk about things orbiting each other leading to mayhem.

There's another kind of nova (ahem, cataclysmic variable stars), known as a *luminous red nova*. It has nothing to do with white dwarves, but they're still called novae, because of the aforementioned historical attempts to categorize stuff blowing up throughout the universe.

Luminous red novae (and because everything in astronomy and astrophysics just has to have an acronym, LRN) happen when two stars crash into each other. You might think this is a pretty common event, what with all the hundreds of billions of stars constantly swirling about the galaxy, but it's actually extremely rare. Yes, there are heaps of stars out there, but there's also buckets of space. Raw, empty space. It's quite challenging to get two stars to even get within a few light-years of each other, let alone close enough to start spiraling in toward each other.

These kinds of collisions probably happen from just the wrong kind of binary pair—an unlucky couple who, through no fault of their own, ends up getting rid of all the angular momentum that keeps them safely separated from each other, dragging themselves into an ever-tighter, and eventually fatal, embrace.[13]

As you might imagine, this is a pretty nasty event. Two stars the mass of the sun can release enough energy to outshine a typical nova by a few orders of magnitude. All that energy comes from the merger itself. Smash a couple cars together, and a lot of energy is released in the form of loud noises, bent metal, shattered glass, and random car parts thrown all over the road. Smash a couple stars together, and a lot of energy is released in the form of a tremendous flash of light and some star parts thrown all over the system.

But stars aren't the only things that can smash into each other. White dwarves can be just as unlucky.

In binary systems, one star will cycle through its life faster than the other, through sheer random chance of being slightly more massive. And this often leads to the now-familiar setup behind everybody's favorite kind of cataclysmic variable star. But the red giant that's feeding the nova outbursts isn't done doing its stellar thing. It too can progress all the way to a planetary nebula surrounded by newly revealed white dwarf (assuming it survives the nova outbursts of its companion intact, which is no small feat).

The system is now left with two white dwarfs, which for various and poorly understood reasons, can spiral in closer together, eventually consuming each other in a final deadly kiss. These mergers release a potent cocktail of elements, especially an abundance of radioactive variants of titanium, which quickly go on to decay into scandium and on to calcium, which is stable enough to stop the decay train and hang on to existence for a while.

Do I even need to say that you don't want to be a part of that whole mess? Not only do you have to deal with the resulting insanity from the collision of two massive white dwarves, but you also have to handle the nuclear fallout and contamination of the system.

That decay process, from titanium to scandium to calcium, also releases an abundance of positrons.[14]

For those not in the know, a positron is a kind of antimatter. And if that hasn't set off alarm bells yet, it should. Antimatter is just like normal matter, which the exact same properties except for one crucial difference: it has opposite charge.

So a positron has the exact same mass and spin as an electron, but is positively charged. That's cute.

What's not so cute is that when matter and antimatter meet, they annihilate each other, converting all of their mass into pure energy. Radiation.

It's the ultimate expression of Einstein's $E=mc^2$, which tells us how much energy is in every bit of mass. In a matter-antimatter meetup, 100 percent of that juicy m is converted into raw, terrible E. To give you a sense of just how senseless this is, two pounds of antimatter, when merely *placed in contact* with two pounds of normal matter in the worst mashup in history, would release the same amount of energy as all the weapons used throughout the entirety of World War II (including the nuclear bombs). Ten times over.

So yeah, positrons. Unless you have some sort of strong magnetic field to deflect them before they can touch normal matter (which is, to be specific, you), if you get flooded with them, you're toast. And if you see two white dwarves spiraling in toward each other in a deadly dance, back away. Back far away. If you manage to survive the blast itself, the fallout just might do you in anyway.

You did pack your Geiger counter, right?

PART THREE

INTERGALACTIC THREATS

Supernovae

The end of my life
Near the end of my days
One more blast,
to show them who's boss.
                    —Rhyme of the Ancient Astronomer

Around 1000 CE, southwestern United States. It wasn't the United States back then, but you get the idea. You've got a pretty cushy job: you're the top astronomer for the local Chaco tribe. Your tribe probably didn't call themselves the Chaco—archeologists would give that name to you hundreds of years later—but we'll call you the Chaco tribe because we don't have any better ideas.

That's right about your job: top astronomer. The Chaco weren't hunters and gatherers. They had figured out agriculture, the trick of bringing the food to you. And like any other farming civilization, they really *really* cared about the time of year. *Is it time to plant yet? Is it time to plant yet? Is it time to plant yet?* They would ask their astronomer for six months straight every single year.

So to set them at ease, you would get up before dawn, hike out to your favorite observing post—away from all the light pollution from the cooking fires—and check out the stars, waiting for the right alignments to signal the proper time in spring to begin the planting season. Job done, you go back to bed. *(No, it's not time to plant yet.)* What else are you going to do? Your only skill is reading the sky. Like I said, cushy job.

So one day you're out doing your sky-reading thing. Moon? Crescent. Venus? Visible. Blazing new star that's so bright it can be seen during the day? Check.

Wait—hold up. Blazing new star that's so bright it can be seen during the day? That's not on the list.

Like most other cultures of the era, when something new shows up in the sky, like a comet or a new star, you flip out. Sometimes folks flipped out in a good way—"Hey, maybe a new king is born!"—and sometimes folks flipped out in a bad way—"Hey, the end times are here!"

We don't know which way you flipped, or what the fallout in your civilization was upon the discovery of that stranger in the sky, but you were concerned enough that you thought the event merited recording. You grabbed your paint and made some graffiti on the nearest cliff overhang: your handprint, the crescent moon, the new star.

That paint baked itself onto the clay in the desert heat, leaving behind a petrograph, still visible hundreds of years later. Your message to the future has been heard. Whatever you saw, and whatever you thought, it was remarkable.

Fast-forward to the present day, and that graffiti tells us what that unknown Chaco astronomer was looking at. With the tools of modern astronomy and our precise knowledge of the motions of heavenly objects, we can rewind the clock. We know that the Chaco lived in modern-day New Mexico around a thousand years ago. The alignment of the handprint gives us an orientation, and the position of the crescent moon gives us a specific date.

While it's not perfect, we have a guess as to what that ancient astronomer saw. We can take the position of the strange star, relative to the crescent moon, and run the clock forward to the present time. And when we point our telescopes into that patch of the sky, we see a familiar sight: the Crab Nebula.[1]

It seems that Chinese astronomers had noticed this "new star" too, which appeared in 1054 (Tycho Brahe wasn't the first person to find a *nova*, he was just the first to write about them profusely and also conclusively prove that whatever they were, they were too far away to be an atmospheric phenomenon). By that time, a thousand years ago, the

Chinese were used to that sort of thing (their bureaucratic record-keeping of the sky already went back thousands of years), so they didn't make that big of a fuss about it. They made a note of the appearance of a "guest star" in their sky and went on to business—according to their traditions, such a new star was a sign that the emperor was misbehaving, so someone had to be the unfortunate one to report it to the man himself.[2] Surely other cultures around the world saw it too, though no European records of the new star exist; although, to their credit, that was an unusually cloudy year for them.

But cultures on opposite sides of the planet noticed and managed to have their record of the sighting persist to the present day. One day, without warning and without anyone asking, a new star appeared in the sky, so bright that it outshone Venus and Sirius. So bright that it could be seen during the day.

Think about it: *during the day.* There's usually only one star you see during the day. You know, the sun. It's so bright that it washes everything else out. But not a supernova, the explosion of a massive star. An explosion so powerful it competes with our own home star for brightness in our Earth's sky.

They don't last long. Every few hundred years a bright new star will appear, far brighter than any nova, visible to the naked eye for a few weeks before slinking back into the void from whence it came. Sometimes they come out of nowhere, born from a previously blank spot of sky. Sometimes they're born from an existing star, and after the flare the star itself disappears from our view.

These are the most powerful single events in the cosmos. These are the last cries of a massive star as it dies. These are the supernovae.

It is now an appropriate time to make like your ancestors and flip out. They tried to warn you, after all.

✪

The Earth astronomer Fritz Zwicky gave them their name, because of course he did. *Nova* is Latin for *new*, and *super* is Latin for *super*.[3] Good, glad we could clear up that definition.

The *super* superlative is definitely deserved: these blasts outshine entire galaxies, their light beaming across the entire universe. They will expend more energy in a week than the sun will over its entire life. Their shock waves stretch for light-years, and their radiation blast goes even farther. Thankfully they're rare, otherwise each galaxy would be swimming in so much radiation that life on any planet would be impossible.

To give you a sense of just how radically bright these things are, consider Betelgeuse, the red giant star on the shoulder of the constellation Orion. It's going to go supernova any day now (where in astronomy "any day now" means within the next few million years). When it finally detonates, it will shine brightly in the daytime sky, just like that new star did a thousand years ago. It will appear as a single point of intense light, and at night it will be brighter than the full moon.

When Betelgeuse dies, it will be so bright that you could comfortably read a book to it at midnight.

And you can stand, in the middle of the night, and watch your shadow cast by the light released during the final furious moments of a giant star.

I'm not exaggerating. Don't believe me? I'm devoting a whole chapter to their physics, so that at the end of it you'll believe me about just how dangerous they are.

You see, it's another death thing. Stars just insist on going out in spectacular ways. Show-offs. Can't anything in this universe just fade away without causing a scene? Well, yes, plenty of things can, like red dwarfs and planets, but they're not particularly dangerous, so we don't worry about them.

I've already talked the death of stars like the sun, and what that means for the death of you—red giant, fits and starts of fusion, planetary nebula, dissolving of the planets, etc. But just like people, there are multiple ways for a star to die. Unlike people, the fate of a star is almost entirely preordained at the moment of its birth. Small star? You will live a long, simple life, burning quietly for trillions of years before fizzling out. Medium star? Near the end of your days you will grow red and violent before turning inside out. A white dwarf is your ultimate fate. Massive star? You burn too hot. You will die soon, leaving only a neutron star or a black hole.

And when those massive stars go, they go with a bang. Don't think you can't ignore them.

While supernovae certainly happen in our own galaxy, they're (thankfully) rare enough that you have to reach truly intergalactic distances before encountering them on a regular enough basis to consider them a hazard. In the Milky Way, a few supernovae go off every century—but they have to be near enough to be seen during the daytime, which only happens every five hundred years or so.

But the universe is a big place, with around two trillion galaxies in our observable volume alone. So essentially, supernovae are happening continuously, nonstop, somewhere in the universe.[4] And since the most powerful explosions can outshine entire galaxies, and it's almost impossible to map and chart every single star in the galactic destination of your choice, the odds are not in your favor that you'll reach a galaxy without some nasty surprise.

As we'll see, stars give off some very distinct warning signs before exploding, so you should be able to stay well clear of them. At the very least, you should simply avoid the most massive stars altogether, as they're already pumping out painfully high levels of X-ray radiation and are usually home to other ferocious energies. This is all *without* the whole blowing up business, which they are prone to do.

✪

Enough fooling around; let's blow something up.

It's a good thing I already told you what happens when a star runs out of hydrogen to burn in its core: it's left with a big fat lump of helium in the center, with a new round of hydrogen fusing going on in a layer around it. Eventually the core squishes down enough to ignite the helium, fusing it into a rock of carbon and oxygen at the center. Yep, it's a good thing I already told you that, so I won't have to repeat myself.

All that business led to a star like our sun getting fat and red and angry, like an evil version of Santa Claus. But what if the star was bigger? What if there was still enough pressure from all that weight to crush down the carbon core?

The answer is that the party keeps on going, only the music changes.

With enough weight pressing down on the core, the carbon can fuse into neon and a few friends. Remember that our sun burns hydrogen at a temperature of around 15 million kelvin. In order to switch to helium fusion, it has to reach ten times that. For carbon ignition? Ten times *that*, or over a billion kelvin. Only the most massive stars, well over eight times the mass of the sun, can accomplish this fantastic feat. Otherwise the carbon and oxygen just stay there, being inert and degenerate, while the rest of the star tears itself to shreds.[5]

But at a billion kelvin, after the star has burned through both its available hydrogen and helium, it can extract some energy from the carbon in its heart.

This won't last long. The helium-burning stage was far shorter than the main sequence lifestyle of the hydrogen fusion, because helium requires higher temperatures to fuse and releases less energy with every event—it has to work ten times harder to fight against the weight of gravity. Carbon fusion burns even hotter, releases even less energy, and lasts even less time.

A massive star can last a few million years along the main sequence (far less time than something like our sun, through the sheer force of its immense weight). Yeah, even the longest stage is only tens of millions of years, not the billions of years we give to stars like the sun. Big stars, man, burning at both ends. Hopped up, no direction, they crash out too fast. Better to live a short famous life than the quiet one of a nobody, I guess.

It will burn helium and produce carbon for a few hundred thousand years, a million tops.

It will eat through its core of carbon in less than a thousand years.

The carbon itself leaves behind an ash: neon. The temperatures climb, the neon fuses.

With so little energy extracted from the fusion of neon into oxygen, after only a few years it's completely depleted from the core.

At a temperature of 2 billion kelvin, the oxygen there burns.

Six months after the onset of oxygen fusion, it peters out.

The fusion of oxygen produces silicon, which starts and ends fusion in a few months.

The very last element produced in the heart of a massive star, created from the fusion of silicon, is nickel. The kind of nickel that's produced is rather unstable; it doesn't have the right combination of neutrons and protons. As soon as it forms it says, "Whoops, never mind, I was never invited anyway," and decays into iron.

More than an entire sun's worth of iron is manufactured in a handful of days.

At each one of these stages the party is getting a little hotter, the music getting a little faster. The core is getting smaller and smaller, since there's less and less stuff to fuse, but the entire star is still bearing down on it. At the same time, each new stage releases less energy than the stage before. Neon + neon to oxygen just doesn't have the oomph that hydrogen + hydrogen to helium did. The end result is that each stage is desperately trying to hold the star up from collapse, working harder and harder, but it's a losing battle.

And just as when we looked at sunlike stars and their core of carbon surrounded by a shell of helium fusion, with that surrounded by a shell of burning hydrogen, these massive stars manage to create hell's own seven-layer bean dip: an inert core of iron, surrounded by burning silicon, surrounded by burning oxygen, surrounded by burning neon, surrounded by burning carbon, surrounded by burning helium, surrounded by burning hydrogen, surrounded by the atmosphere of the rest of the star.

All these layers in constant flux and turmoil as fusion reactions start and stop and trade places.

Finally, the iron core begins to fuse. But at this point, it's futile. At every one of the above stages the fusion process released energy—that's what was keeping the star going, after all. But at each new stage we got less and less energy out of the deal. And iron isn't a producer of energy, it's a vampire. The overwhelming gravity of the surrounding mass of star does fuse iron into heavier elements, but it *spends* energy to achieve that fusion.

For millions of years, even in these final heady days, there's always something to counteract gravity. There's always an explosive release of atomic energy, putting up the good fight and resisting the never-ending crushing onslaught of the star's own weight.

But no more. There's no new source of energy. The end is here.

So there we are: an inert lump of iron, a pit in the stomach of the lion. At first, nothing happens. The music has stopped, but the partiers are still there, awkwardly staring at each other, waiting for the beat to come back to life.

This, like all things, can't go on forever. The last moments of a massive star's life, after millions of years of steady and stately burning, is a total chaotic mess of some of the most extreme forces ever to be seen in the universe.

At first this iron core is stable, held up by the same *degeneracy pressure* that gave birth to the white dwarf at the end of the life of smaller stars. It continues to heat up and contract, as more and more iron rains down on it from the hellish shell of silicon fusion above it.

It looks like, for a brief moment, that the massive star will end its life the same way that its younger cousins did: by fitfully and fatally turning itself inside out in a gory dance of death. Perhaps a larger version of a planetary nebula is on the horizon?

But within fifteen minutes of the appearance of an iron core in the heart of a massive star, it will be all over, violently.

Remember how I briefly mentioned that degeneracy pressure can only go so far? Yes, the quirks of quantum mechanics offer a source of pressure able to stave off complete gravitational catastrophe. But even quantum mechanics has a limit; the pressure supplied by the electrons in the iron core can crack. Eventually too much iron settles into the core, overwhelming the degeneracy pressure that was briefly holding up the star. Then things so haywire, *fast*.

Once the support from electron degeneracy fails, the core simply collapses, so quickly—parts of it reaching a quarter of the speed of light—that it completely detaches itself from the rest of the star. Like the pillars knocked out of a skyscraper, the rest of the star follows blindly along, trying to chase after the core that supported it for so long. There's nothing stopping it: the only force is gravity, which has finally won out against the energy produced in the core.

As the core shrinks and folds in on itself, it reaches truly unimaginable densities, far surpassing even the crushing extremes of a white dwarf. It becomes so intense that any stray electrons find themselves *literally inside atomic nuclei*, a place where they absolutely hate to be. And once there, they then find themselves crammed inside of protons, their very mortal enemies.

When a proton is squeezed together with an electron, two things happen. First, the proton becomes a neutron. Second, the reaction produces a neutrino. Remember those little buggers? Tiny particles, barely with any mass at all, streaming through you right now, but you don't even notice. I called them the Least Threatening Particle, Ever, but maybe I was wrong. Oh, they won't hurt you directly, but they'll still find a way to boil your blood and bones. You'll see soon enough.

One by one, iron atoms melt from the assault, and their protons swallow electrons, turning into neutrons in the process.

The end result is a giant, ghastly core of neutrons, weighing more than the sun but compressed down into the size of a city, surrounded by layers upon turbulent layers of infalling stellar atmosphere.

And this core can, at least for a moment, resist the continued breakdown of the star.

You can think of electron degeneracy pressure like the roomful of dancers at a party. You can only put so many people in the room and still give them room to dance. Try to add more people, and they'll complain and kick them out—offering a source of pressure support against gravity. They need their room to groove, after all.

But if you succeed in shoving more people through the door than they can kick out of the party (the collapse of the iron core from the loss of electron degeneracy pressure), the dancing breaks down, and everyone starts to squeeze together. But then you run into another limit: you can only cram so many *people* into a room, whether they're dancing or not. Cheek to jowl, floor to ceiling, there are only so many bodies that will fit into that room.

For a star, that second limit is reached when the density of the core reaches that of an atomic nucleus. So the core is no longer just a dense clump of iron atoms; it's now essentially a *single atom a few miles across*: a giant ball of neutrons all shoved together as close as they can get.

Those neutrons naturally repel each other. Not through any electric repulsion, since they are, after all, neutral, but through a combination of the vagaries of the strong nuclear force and their own version of degeneracy pressure. Neutrons are fermions too, which means they can only be squeezed down so tightly before they refuse to go any further. And since neutrons are much, much heavier than electrons (almost two thousand times heavier), they can cram much more tightly together before the wonderful quantum degeneracy effects kick in.

But kick in they do, and that city-sized atomic nucleus made of almost entirely neutrons stops its collapse, looks outward at the crashing tsunami of plasma rushing into the core, and says "not today."

Here comes the bounce.

In less than a second the star goes from "I don't see anything wrong with it," to "What star? I don't see a star. Just a massive explosion headed right fo—"

All the material surrounding the core was rushing in at a solid fraction of the speed of light. I said when the party's over, the party's over. Hooray for gravity! But all this *stuff* slams head-on into that solid lump of neutrons. And they're not budging, no way, no how.

A giant star's worth of material. Moving fast. Hitting an impenetrable wall.

All that gravitational energy, all that pent-up frustration, can now only go in one direction: out.

Now here's where it gets tricky. You see, we don't know *exactly* how a supernova actually . . . supernovas. We know that there's more than enough built-up gravitational potential energy to power the big boom, and we know the basic sketch of the process. But as hard as we try in computer simulations, we just can't get a collapsed star to blow up. Right *after* it starts to blow up, we're golden: we can match observations just fine. But the blowing-up moment itself? It's complicated.

It's complicated, but here's the basic best guess idea. It's important that you know this, because if supernovae go down this way, it means you have some potential ways for detecting one before the big blast hits you.

Where were we? Right, the bounce. Tons—or more accurately, *suns' worth of tons*—of material ricochet off the core, expanding outward in a shock wave. A sonic boom of star stuff.

But then it . . . stalls. The rapidly expanding shockwave of fury quickly peters out. Once it does so, we're in a slightly awkward position: we're left with a dense core, a giant star's worth of material surrounding it, and a wannabe explosion that's rapidly losing steam in the middle of the atmosphere.

Neutrinos to the rescue.[6]

If you'll remember, during the calamitous collapse of the core, it's a ball of hot mess. And in that hot mess, nuclear reactions occur at an insanely high rate, converting protons into neutrons. Each one of these reactions produces a neutrino as a byproduct. Neutrinos happen as a result of all nuclear reactions, so they were being created all along during the star's lifetime, but now it's serious.

Over 99 percent of the energy released during a supernova event isn't even seen, and can't even be seen, because it goes into a giant nuclear neutrino factory, a tremendous burst lasting less than a second. Almost all the neutrinos just stream away from the star—they hardly ever talk to normal matter, so once they're gone they're gone.

But a small fraction of them—maybe even 1 percent—do occasionally hit something in the star before escaping. The shock blast wave, slowing down in the outskirts of the star, gets a tiny little kick. And another. And another. And several trillion more.

And away it goes: reinvigorated by the neutrino blast, the shock wave accelerates out of the star, carrying more and more material with it. Or not.

Like I said, the details are a bit sketchy. Many times in our fancy computer simulations of this momentous event, even powered by neutrinos, the shock wave still stalls again. Lame, I know. We have a hard time generating the exact sequence of events during the last second of a massive star's life, but it seems that an incredible amount of sloshing occurs, driving the guts of the star back and forth in an insane resonant pattern, focusing the energies needed to finally make breakout.[7]

But once that shockwave finally makes it to the outer surface of the star and into space . . . well, boom. An entire star—much more massive than the sun—exploded out across the local region of space, in less than a second.

There are reasons you don't want to be near a supernova when it ignites. Do I really need to spell it out for you? Probably not, because you're pretty smart, despite what your mother says about you. But I'm going to anyway, because I really enjoy talking about all this.

In fact, you don't want to be anywhere near a giant star before it goes kablooey in the first place. At the very end stages of their evolution, when their cores are developing that nasty onion layer of various combinations of fusion reactions, the surface swells, expands, and heats. After all, if you had at least seven varieties of nuclear fusion happening inside of you, you might start to get a little bit flustered.

As they continue to evolve and age, the surfaces reach scorching temperatures, easily topping tens of thousands of kelvin, and sometimes hitting highs in the six figures. It's so hot that the copious UV radiation (the same kind that give you suntans, sunburns, and carcinoma) is able to mess with the molecular makeup up the surrounding gas, giving them perplexing spectral lines that were a mystery to astronomers for decades.

These kinds of stars, known as Wolf-Rayet stars in honor of their mid–18th century discoverers (who basically just noted that something funky was happening with their spectra), are gargantuan, swollen monsters, reaching over twenty times the width of the sun and a brightness well in excess of millions of times what our home star can produce.[8]

That intense radiation, and an insane stellar wind to match it, drives away a layer of the atmosphere, making a nebula around the fitful soon-to-be-dead monster at the center; a forewarning of what will come.

And this is all *before* the big explosion.

At the moment the shockwave from a core-collapse supernova makes its way to the surface, an intense radiation blast flashes across the system. Indeed, it flashes across the entire universe. The phrase "outshines an entire galaxy" sounds very poetic in an academic context. It sounds quite different if you're within, oh, say, a light-year of a supernova.

Think of the heat of the sun on the planet Earth. Think of a hot summer day, the rays beating down on you. No respite, even in the shade. Stand in that spot, sweat running down your forehead, the skin on the back of your neck tingling with the beginnings of a sunburn.

Stand there for eight billion years. The entire lifetime of the sun. Not moving, absorbing every bit of solar radiation you can. Every photon. Every scrap of heat. Every day, for eons.

For a supernova, take all that energy, all that burning, all that intensity adding up over the years, but unleash it in a few days.

That's the kind of radiation I'm talking about.

And that's just the light, the photons, from the breakout. Then comes the blast wave itself: a tsunami of radioactive particles, accelerated to 10 percent of the speed of light, a supersonic blast wave ripping through space, heating up anything and everything it touches.

This shockwave is a wall of neutrons and protons, slamming into each other and the remnants of the star that have already been ejected. These particles are traveling so fast that there's enough energy to fuse elements heavier than iron. This couldn't happen very easily even inside the largest stars, because the process drains energy, but once the star's blown up, who cares! Let's spend energy like it's our last day in the universe, because it is.

New elements: xenon, gold, lead, potassium, and more. All are manufactured in the insane cacophony of a supernova blast. It's one of the few places in the universe with the right combination of ingredients and temperatures to make this happen. It's called *nucleosynthesis*, if you're keeping score.[9]

Imagine standing inside a nuclear reactor. A reactor exploding at the rate of tens of thousands of miles per hour. Your body, your ship, your everything-you-know would be reduced to a cloud of slightly radioactive space dust.

Some of the material gets swept up, trapped in the shock wave, rattled around, and ejected at a speed close to that of light. This mix of protons and heavier nuclei, now called *cosmic rays*, go on to stream throughout the universe, soaking unsuspecting travelers with poison from the blast.

The radioactive blast wave continues to expand, stretching light-years across, sweeping up material as it goes. And as the days go by after the initial explosion of the star, it actually gets hotter. Some of the elements created in the maelstrom are radioactive, and they decay, quickly. These decays produce even more heat, adding fuel to the fire, increasing the radiation output, causing the expanding shell of debris to glow ever more fiercely.

At its peak, about a week after the initial deadly explosion, the ejected supernova material will become ten times brighter than the initial shock wave breakout, about a billion times brighter than our sun.

Eventually, though, the system runs out of radioactively decaying material to power itself, and it loses steam, cools off, dispersing like dust in the wind. But the danger zone persists for up to a hundred years—the radioactive elements with longer half-lives continue to dance in the wreckage.

Take for instance the Crab Nebula, the result of the awe-inspiring supernova of 1054. Even a thousand years later, with the guts of the former star spread out over a volume ten light-years across, still maintains a roasting 15,000 kelvin, and is home to a rich, potent soup of radioactivity. The magnetic fields in the nebula alone are strong enough to whip up particles to half the speed of light.

A Chernobyl-like Exclusion Zone full of radioactive particles and intense X-rays. No sane traveler should enter.

✪

There are other ways to make a scene. Sometimes, a do-nothing middle-of-the-road star gets a chance to see its name in lights.

The story of death and destruction I just told—called *core collapse* if you're a normal human being and *Type II* if you're an astronomy nerd—was based on the life of a giant star in isolation; this is simply how they end their lives. But stars are hardly ever alone. The sun has a few planets to call its friends, and many stars—about half—come in pairs, or triplets, or even more.

And interesting things can happen in your life if you have a friend.

We met in the last chapter a common sight in every galaxy: a pair of stars, one already in its post-death stage as a white dwarf, and the companion in its near-death stage as a red giant. With the right setup, some of the atmosphere from that giant can be siphoned off onto the surface of the white dwarf, creating a thick blanketing atmosphere of hydrogen.

When things go critical in that layer of hydrogen, it flashes into a nova, a flare of light that dazzles and awes, burning through that hydrogen in a fantastic thermonuclear event.

But it can get worse. Much, much worse. Like when that layer of gas from the red giant companion cracks the white dwarf like a shell of an egg.

Sometimes the hydrogen can pile up without ever having the exact right conditions to fuse itself. It just gets thicker. And heavier. Inside that layer, the white dwarf starts to react. The pressures and temperatures climb. As the white dwarf nears the limit of stability, as gas piles up on the outer layers, the core of the white dwarf starts to convect, just like the plasma inside the sun and just like a boiling pot of soup. A soup of oxygen and carbon, not exactly tasty, but viscerally intense.

This isn't a stable situation: a planet-sized nuclear bomb, weighing as much as the sun, just waiting for a fuse to set it off.

Somewhere, at some point on the white dwarf, the temperature and pressure from the thick hydrogen atmosphere get high enough to ignite the fusion of carbon. That little spark, that tiny fire of fusion, releases a little bit of energy. The released energy is enough to trigger carbon fusion in the area around it, which in turn spreads the fusion reaction even further.

Completely without warning, a crisis point is reached. A spark has been ignited inside the white dwarf, a brief flicker of fusion.

And it spreads throughout the star, like the lick of a flame on a puddle of gasoline, igniting fusion wherever it goes. In a normal everyday star, this process would be confined to the core, keeping the temperature and pressure regulated so that it didn't get out of control. In those cores, the increased heat from the fusion would puff out the star, easing the pressure and cooling off the rate of reactions. But a white dwarf isn't a normal star: it has nothing to balance out the increasing temperatures and the ever-widening flame front.

It's a runaway event. More fusion means more heat, which means more fusion, which means more heat, which means more fusion.

In just a few seconds, the flame goes from calm-steady-watch-the-pretty-glow to oh-snap-it's-exploding-the-whole-star.

Here's a word you can import into your everyday life: *unbind*. You're bound together: the atoms and molecules that make up your body are kept together with a variety of forces. With enough energy, say, a grenade, I can separate the molecules in your body and send them flying apart in every direction. You would be *unbound*. Also, dead. Since this process takes a lot of energy, it doesn't happen very often and is generally considered extremely violent.

It takes a heck of a lot of energy to unbind a white dwarf. They have enough mass to build a star, compressed into something no bigger than a rocky planet. Think how much energy it would take to pluck every single atom out of the white dwarf and give it enough of a boost to reach escape velocity. It's a lot of energy, and so white dwarves can maintain their internal composure for trillions upon trillions of years.

But an *uncontrolled nuclear fusion explosion* will do the trick.[10]

Before you know it, the white dwarf has unbound right into your face.

Shock waves comparable to the speed of light. Flares that outshine ten billion suns. You know how it goes: a supernova.

By the way, these are called *Type I supernovae*. Yes, I talked about them after *Type II*, deal with it. As to why they have these names, you can once again blame Fritz Zwicky, who I guess thought this would be useful. We can't judge him too harshly though: a lot of this nomenclature was worked out *before* we knew what was actually happening. This is a state of affairs frustratingly common amongst astronomers, who have a tendency to classify and name things based on what they see in their telescopes (which is almost always various kinds of elemental spectral lines), and decades or centuries later the physicists catch up to explain the observations, but by then the names have stuck.

These Type 1 supernovae occur almost as often as the core collapse variety, but they are arguably worse. At least with Type II, you can spot a giant star in its death throes and have some warning. With Type 1, it's just another run-of-the-mill binary situation of a white dwarf partnered with a distended red giant, the gas swirling around its accretion disk on its way to the surface of the smaller of the pair.

It's a pretty sight, attracting astronomers and visitors alike.

And then, before you even know it, the whole thing implodes, and the star is vaporized in a flash of light ten times brighter than a core collapse supernova.

The companion star—if it survives the blast at all and isn't just ripped to shreds—can get its outer layers stripped away and booted out of the system, sent on a rogue trajectory through the galaxy. So much for a lifelong partnership.

✪

And just when you thought it couldn't get worse than a supernova, nature just has to go and show off. There are a few special forms of supernovae that are especially nasty, as if the *regular* forms of supernovae weren't nasty enough.

The terminology here is a little confusing, because the astronomers trying to understand these explosions are having a hard time understanding these explosions. So over the decades various names have come into and gone out of fashion; the bell-bottoms of the astronomical community.

What's more, since we only understand these kinds of extreme events by watching them from a safe distance (nobody, even the foolhardiest adventurer, dares to go for a close-up look), and these events are understandably rare, we don't have a lot of observations to go on.

One name that gets tossed around is *hypernova*, because that's the next logical progression from *super*. Another term is *superluminous supernova*, because that's also the next logical progression from *super*. The brightest of such supernovae found to date shone brighter than 570 billion suns, or twenty times that of the entire Milky Way galaxy.

Generally nowadays hypernova refers to an explosion where the shock wave itself is going especially fast, while the superluminous supernovae are . . . just really, really bright.

Another term that's hung around for decades is *gamma ray bursts* (or, GRBs, but definitely not "gerbies"). These were first spotted in the 1960s when the US military wanted to snoop on Soviet nuclear bomb tests, which emit copious amounts of gamma ray radiation, and were surprised to find extremely energetic gamma rays coming from all over the place, but in the opposite direction from where they expected them.

I'll tell it to you straight. We're not exactly sure what causes hypernovae. We're not exactly sure what causes superluminous supernovae. We're not exactly sure what causes gamma ray bursts. We're not exactly sure if they're all related, different varieties of the same thing, completely different, some amount of overlap, or what. But whatever is causing them, they take the phrase "off the charts" literally to a whole new level—astronomers are still busy inventing whole new charts just to catalog them.[11]

When it comes to gamma ray bursts, it's thought that at least some of them are related to peculiar supernova explosions. During the final collapse phase of a supernova, there's a lot of energy released (which is a pleasantly euphemistic way of saying "exploded"). Usually that energy is distributed evenly throughout the exploding star, making it go boom.

But the star is a) big enough, b) rapidly spinning, and c) some other vague set of poorly understood conditions are just right, then as the material from the star collapses in a headlong rush to the giant ball of neutrons that was once called the core, the material can be funneled into two powerful jets that race away on either end of the star. It's the same kind of physics that generates the potent jets around newly forming stars and around black holes. Physics is physics the universe around, and it turns out that nature doesn't have a lot of tricks in her repertoire.

Supernovae are already some of the most energetic events in the universe, but with this type of GRB a lot of that energy gets funneled into two narrow streams, rather than blasting off in every direction. If you happen to stand in the path of one of these jets, good luck keeping your hat on, because you just got fire hosed with gamma rays, by far the deadliest form of electromagnetic radiation around. Think of how much more dangerous X-rays are compared to, say, a soft household lightbulb. Gamma rays make X-rays look positively puny.

Even planets can't withstand their onslaught: if a burst happened too close to Earth (as in, within the same half of the galaxy) and it was aimed directly for the solar system, a dose of only ten seconds would overwhelm the atmosphere by eating away at the protective ozone layer. Without the layer, life on the surface gets irradiated by the usual rain of cosmic rays. It's even suspected that an extinction event on Earth 450 million years ago may have been ignited by a trigger-happy GRB close to home.

But funneling into jets isn't the only way a supernova can go haywire (as if supernovae themselves weren't already the very definition of "haywire"). In the very most massive stars (we're talking well over a hundred times the mass of the sun), the bottom can fall out before the party even gets going. The culprit is gamma rays (Are you starting to see a connection between the most energetic events in the universe and the deadliest form of radiation out there?), and it might be a possible cause behind the superluminous supernovae.

Nuclear reactions in a star produce high-energy radiation. When the temperatures in the nuclear core skyrocket, that radiation shifts to higher and higher frequencies. In truly giant stars, the nuclear furnace starts to produce gamma rays. Which doesn't sound so bad at first, except that gamma rays can play a special trick.

Extremely high gamma rays can turn themselves into matter. It's not a magic trick; it's the same voodoo of particle physics that drives showers of cosmic rays. A gamma ray can, on a whim, decide to become a pair, an electron and its antimatter counterpart a positron. Now sometimes, under normal conditions, the electron and positron find each other again and annihilate, returning back into a gamma ray.

But the core of a colossal star is not a normal condition. When things go haywire, loose positrons and electrons aren't able to find each other and annihilate, sapping energy from the core as gamma rays simply blink out of existence. When this process starts to happen in abundance, the star loses one of its main sources of pressure support: the raw radiation streaming out of the core. The star then collapses a bit, which drives up the temperatures, which produces more gamma rays. But those gamma rays are lost to more pair-production of electrons and positrons, which makes the star collapse, which drives up the temperatures . . .

It's a story told over and over again throughout the universe. Once equilibrium is thrown out of whack, it's hard to get back to any sense of normalcy, and the result is usually disastrous.

If a star is over 130 times the mass of the sun, this process is unstoppable.[12] The core contracts so quickly and so completely, with the temperatures reaching unthinkable levels, that all material in the core vaporizes in a single nuclear fireball, like the helium flash inside of red giants, but so much worse. That momentous release of energy is more than enough to unbind the entire star.

In a moment, the entire star explodes from the inside out. The single greatest nuclear explosion known to the universe.

At least 130 suns' worth of material, blown clean away. No neutron core. No bounce. No neutrinos. No complicated dance of obscure physics. Just raw power; nature's ultimate fury released in a single event.

Sure, GRBs, hypernovae, and superluminous supernovae are extremely, frighteningly dangerous. But what the heck are you doing near such massive stars in the first place?

✪

You've got to be careful out there. Core collapse supernovae come from only the most massive stars, which tend to form in the spiral arms of galaxies, where increased star-formation rates produce too many massive short-lived stars. If you stay out of the arms, you should be safer.

Watch out for massive stars near the ends of their lives. If the star is more than ten times the mass of the sun, you're at risk of getting too close to an eruption. The entire transition from normal-looking star to explosive death blast takes only a few seconds, with very little warning. Stay clear—at least tens of light-years away, where the damage from the shock wave and gamma rays should be minimal.

But Type I supernovae come from white dwarfs, the leftover husks of normal medium-sized stars, like the sun. They're everywhere, in the spiral arms, in the disk, in the bulge, even in the small globular clusters orbiting each galaxy. They're sneaky too: without detailed observations, it's difficult to tell if a white dwarf is accreting material from a companion. You put yourself at risk by getting close enough to study it. Avoid multiple-star systems, unless you are positive that all stars in the system are in their normal hydrogen-burning phase.

The neutrinos are a warning. They are launched early in the event, and aside from kicking the shock wave in the rear end, most just stream right through, escaping the cacophony. That means that the neutrinos will reach you before the light (and radiation and cosmic rays and death) do. If you have a well-tuned, direction-sensitive neutrino telescope, you can maybe buy yourself a few seconds' warning.

Recent supernovae are best avoided too: there's enough high-intensity radiation to make the surrounding cloud of gas glow for a few thousand years, so what do you think they would do to your body?

Fortunately, supernovae are relatively rare: just a few per galaxy per century. And if you're parked in a relatively calm neighborhood, with no obvious candidates lurking around, you're probably good for a permanent settlement. But choose wisely. If you're curious, lists of possible supernovae candidates are available online. This is a warning service provided for free to travelers; use it!

There's a silver lining to these supernovae. They're factories, furnaces that forge oxygen and carbon inside the stars and heavier stuff in the explosion. Look at what you're made of: bones of calcium, blood of oxygen, cells of carbon. You eat your bananas for their potassium, and build your computer chips out of silicon. It was all made is the death throes of a massive star, generations ago. You are the ash, the cinders, the leftovers from those short-lived beacons.

You owe your very existence to supernovae, and you should pay them the proper respect.

Neutron Stars and Magnetars

You're dead but yet you shine
Reminding all
of what you once were.
We will not forget you.
    —Rhyme of the Ancient Astronomer

I'm warning you: to get through the next hazard I'm going to have to talk about pasta. You'll see.

White dwarves may seem strange and dangerous, with their weird almost-crystal interiors and constant stream of hot X-rays. Supported against their own crushing weight by the impossibility of electrons cramming themselves in too tightly, it seems like nothing can be odder.

But this is the universe we're talking about. There's *always* something odder.

White dwarves form from the leftover cores of mid-sized stars. But what about big stars? The ones that cause those massive supernovae explosions? Just before that final explosion that signals to the rest of the universe that a star has died, the electrons get shoved into the hearts of the nuclei of the atoms in the core. There they combine with any protons that they find and turn them into neutrons. Plus neutrinos. There are always neutrinos. And we saw before just how important they were in driving the Big Pop.

So the nuclei, which used to be happy little bundles of protons and neutrons, lose almost all their protons. The dense stellar core transforms from a ball of free electrons swimming around some nuclei to a ball of free electrons swimming around some neutrons.

As with their white dwarf cousins, the game is the same: neutrons can only pack in so tightly before they just *won't*. It's another degeneracy pressure supporting the defunct star against the eternal crush of gravity. But even though the game is the same, the players are different. Neutrons can pack in much more tightly than electrons. In fact, they can reach densities around that of an atomic nucleus before they'll start resisting.

That resistance is what triggers the ultimate explosion, as the surrounding atmosphere races in at a healthy fraction of the speed of light, before careening off the neutron-rich core and detonating.

And after all is said and done, after the brightness that can outshine a galaxy, after the rain of hellfire that can enrich the periodic table, after the shock waves and turbulence and radiation, something can remain. The core that unwittingly triggered the detonation of an entire massive star can withstand the fury, a dense enigma of neutrons and electrons crammed as close as they possibly can be, radiating heat at a sweltering thousand billion kelvin.

An object a few miles across with the density of an atomic nucleus.

More mass than the sun compressed into a volume no bigger than a city.

A dying husk of a once-giant sun, exposed to the vacuum of space through the power of an entire star blasting itself apart. Angry, bitter, vengeful.

Behold, one of nature's most fearsome creations: the neutron star.[1]

Let's explore the densities involved in the making of a neutron star, because not only is this kind of material impossible to re-create in any laboratory, it's downright almost impossible to even contemplate. I mentioned before about how the separating of electrons into different energy states gave everything volume. If you were paying attention—and I hope you were—you would've read into the implication: that without the electrons, everything would be a lot smaller. Almost all of an "atom" is just miserably empty space. The nucleus of protons and neutrons take up only a measly 0.0000000000001 percent of the volume of an atom. Yes, you read that right.

So take that incredibly tiny volume, with its huge densities, and make a star out of *that*.

The numbers involved to describe neutron stars quickly become outrageous. So outrageous that just typing them doesn't get across their outrageousness. If you could reach down and grab a handful of this material, it would weigh more than a thousand Great Pyramids. That dwarfs (pun intended) the same amount of white dwarf material, which would only weigh a few hundred tons. If you dropped it, it would fall back to the surface of the neutron star in less than a microsecond at a speed of, oh, let's say, a few million miles per hour.

Don't trip. Anything falling onto the surface would strike with such speed that it would simply obliterate, turning whatever it was into just more neutron star stuff. You may fall onto a neutron star as a person, but you'll land as a pancake. If you wanted to leave, you would need your rocket to boost you to as much as half the speed of light. That works out to a gravitational pull a couple hundred billion times stronger than the Earth. Not so bad as a black hole, of course, but pretty hellish.

And forget going on an adventurous mountain hike: the biggest "hill" on the surface is no taller than a few millimeters. Take a dozen hairs on the back of your neck, which ought to be standing up by now, and that's the thickness we're talking about.

Neutron stars have such intense gravity that beams of light can be trapped in orbits around them. Which means you can see multiple sides of a neutron star from one vantage point.

The outermost layer is a crust-like material, but one unlike any doughy crust you would find at your local baker. It's made of electrons and free nuclei, held together by the extreme gravity of the rest of the star. Below that the dough analogies continue in a hilarious way.

I should warn you that we're not *exactly* sure what goes on inside a neutron star. It's not like we have any handy experiments to compare against. Ever make a neutron star in a laboratory? Yeah, me neither. But what guides scientists is math and physics, which is all we have to go on.[2]

In the interior of a neutron star, there's a strange balancing act. The pressure of gravity has squeezed apart even atomic nuclei, allowing their bits to float freely. It's mostly neutrons down there, hence the name, but there's also a few surviving protons floating around. Those protons would normally repel each other, being like-minded charges and all, but they are forced close together as the strong nuclear force tries to bunch them up with their fellow neutrons.

This is a complicated dance of physics under extreme conditions, resulting in very odd shapes. It starts near the surface with blobs of a few hundred neutrons that are best described as gnocchi. Below that the neutron blobs glue together into long chains: we have entered the spaghetti layer. Underneath that, at even more extreme pressures, the spaghetti strands fuse side by side and form lasagna sheets. Under it all, even neutron lasagna loses its shape, becoming a uniform mass. But that mass has gaps in it, in the shape of long tubes. At last: penne.

I wish I was making any of this up, but Earth scientists must have been especially hungry as they discovered these structures.

Below that, in the very center of the neutron star, lies . . . well, we don't really know. It's certainly some exotic plasma made of some of the most fundamental particles in the universe. Maybe even neutrons themselves are broken down into their individual quarks.

Whatever it is, I'm sure once it's figured out, there will be a pasta metaphor to describe it.

✪

That's the inside of a neutron star, which I hope you'll never have to personally experience. Well maybe "experience" isn't the right word. "Become a part of," perhaps? But that complicated physics, with the crust and the pasta and all the rest, dictate what happens outside a neutron star, which is where you should care.

Initially neutron stars are very hot, just like their white dwarf cousins. But whereas the dwarves take a long time to cool down, the neutron stars are much more efficient, turning down their furnaces to only a million kelvin in just a few years. Of course, even a million-kelvin star is enough to emit enough X-rays to roast you alive, so they're no slouches.

Neutron stars spin. Oh man, do they spin. If the original star was spinning even a little bit, as it collapses into a neutron star it speeds up something ridiculous. As additional material falls in it only serves to speed it up more, like a deranged carny taking the safeties off the merry-go-round. How fast? Good question. The fastest neutron stars spin around every few milliseconds. Try that on for size: a star twice the mass of the sun, compressed into an area no bigger than Manhattan, spinning a few hundred times every *second*.

That's a quarter of the speed of light.

Fast-spinning neutron stars do slow down . . . eventually. "Eventually" meaning a few million years. So if you're patient, you can outlive them.

But sometimes neutron stars randomly speed up for no apparent reason. Known as "glitches," they are thought to be caused by starquakes on the crust. That thin shell of electrons and other unlucky particles is under an enormous amount of stress, and when the stress builds up too much, it cracks apart and resettles, in the process releasing a fantastic amount of energy, mostly in the form of a flare of gamma rays. For the folks really into their astronomy jargon, these are technically "soft" gamma rays, as in the weaker form of gamma rays, but guess what: when it comes to irradiating you, your body can't really tell the difference.

So we can see neutron stars glitching from the burst of radiation as a starquake ripples through the crust, which should give you some clue as to the amount of raw power entombed in a neutron star: when they *blink* they're almost as deadly as a supernova.[3]

Once the crust resettles, it will be slightly smaller, because whatever inflated and stressed it is now relieved. But not a lot of mass is lost in the glitch event, so by conservation of angular momentum, the neutron star speeds up just a tiny bit, as if it weren't already going fast enough.

The neutron star spin slowly decays over the millennia because their incredibly strong magnetic fields bleed energy into space. Oh, I haven't mentioned the magnetic fields yet? How embarrassing. You know I couldn't let a chapter sit idly by without bringing up magnetic fields, especially strong ones.

I've talked about the strong magnetic fields of the sun. Once we get further into the universe, I'll warn you about the dangerous fields in blazars and quasars. But those are nothing. Two-bit players. Wimps. Chumps. This is the Real Deal: the strongest magnetic fields in the known universe, thought the be generated by the same dynamo mechanisms that power all the other magnetic fields out there. Even the weakest neutron star magnetic fields are magnetic beasts. Check out this number: a hundred million. A hundred million times stronger than the Earth's; that's where neutron star magnetic fields *start*.

These magnetic fields can pick up any nearby bit of material, rip it to shreds, and send it flying back out into the depths without even breaking a sweat. They are such incredible sources of energy that they fuel the creation of their very own nebulae—the *pulsar wind nebula*.

You remember the Crab Nebula, the remnant of a supernova from over a thousand years ago? In the deepest heart of that nebula sits the leftover core of the star that generations of humans reliably saw in their sky night after night. Now it's a zombie, dead yet still bright, blowing a cavity inside the tattered remnants of the stellar atmosphere that once entombed it.

Neutron stars with the most powerful fields get their own name—*magnetars*. No, that's not some super-villain in the latest summer blockbuster. It's just a star with a magnetic field a quadrillion times stronger than the Earth's. That's right: we've skipped past billions and trillions, cranking it up all the way. We're not playing around anymore. With fields this strong, life gets a little bit strange.[4]

Instead of just nudging around a needle on your compass, these fields deform your very atoms. Those little electrons whirring around in their orbits? They respond to magnetic fields, just like any other charged particle. A strong enough field can overwhelm the forces that keep an atom in check, bending it to its own evil will. In the end, you go from nice ball-shaped atoms to . . . cigar-shaped atoms? Needle-shaped atoms? Whatever it is, it's nasty business, and chemistry grinds to a halt.

At just 1,000 miles away, these fields are also strong enough to magnetize your body.

That's not something you normally worry about, is it? You can play with magnets all day long without thinking about their effects on you. Even the strongest human-made magnetic fields don't really affect people. But get near a magnetar or even a regular neutron star (side note: I would've called them *neutrotars*, but nobody listens to me), and your body chemistry simply breaks down. The arrangement of molecules, the biological processes that keep your heart beating and your brain thinking, can't function.

And then you just dissolve.

When those little starquakes briefly rupture the surface of the neutron star and cause it to glitch, the magnetic field can get an unexpected boost. A change in magnetic field leads to an electric field, which leads to a magnetic field, which leads to . . . light. Electromagnetic radiation: the gamma rays.

Besides the occasional flare-up, the strong magnetic fields can also power twin jets of radiation, funneling charged particles to extreme enough velocities to emit high-energy radiation in tight beams. Usually these jets aren't perfectly aligned with the spin of the neutron star—Because why should they be?—giving the biggest and baddest strobe lights in the universe. If you get caught in the beam, you're toast. Nobody likes a laser pointer in the eye, and nobody likes a *pulsar*—the name we gave to these flashes of light—in the face.

From far away though, they're handy little beacons, flashing hundreds of times per second, each one with a unique signature, which is how we first detected them back on Earth in the middle of the 20th century.[5]

Since each pulsar has a unique frequency, they can be used for navigation: find a few known pulsars and you can triangulate your position anywhere in the galaxy. Beep, beep, beep, guiding you through the dark. Just don't get so close.

While only a few thousand have been charted, there may be as many as a hundred million neutron stars in every galaxy. Unfortunately for us we're only able to detect the youngest ones—the ones with the biggest, strongest magnetic fields, or the ones in binary systems. The younger ones are easy to detect—they are brightly shining freaks of nature, after all. But if they're not rotating or accreting, there's nothing to power visible emission. They're just hunks of neutrons and angry gravity, cooling down along with the rest of the universe. While they're not nearly as dangerous, their gravitational grip is almost as bad as black holes.

Stay careful out there, in the dark.

✪

As with white dwarves, be especially wary of neutron stars in binary systems. Thankfully these are much rarer than their carbon + oxygen counterparts, with only an estimated 5 percent of all neutron stars living with a sibling.

But they get vile, quick.

Recall the story of the novae. White dwarf has red giant companion. Red giant's atmosphere pours onto the surface of the white dwarf. Temperatures and pressures reach critical threshold. Best case outcome: intense flash as the hydrogen layer fuses itself into oblivion. Worst case outcome: the entire white dwarf rips itself apart in a nuclear meltdown.

Now take this same scenario and ramp up the densities and energies.

Sometimes material can settle onto the surface of a neutron star from an orbiting giant companion. Sometimes that material can reach a critical threshold of temperatures and pressures. Sometimes that material can ignite in an uncontrolled nuclear event. All the times that happens, a giant burst of X-ray radiation blasts into the universe.

When white dwarves collide, they release enough energy to form new elements that quickly begin radioactively decaying.

When neutron stars collide, they get a whole new name: the *kilonovae*. So designated because they are, roughly, about a thousand times brighter than a typical nova. In the Universal Hierarchy of Explosions, they're not exactly top dog, easily outranked by supernovae and hypernovae, but they're certainly bad enough.

Bad enough that the resulting collision, and accompanying release of all that juicy energy, can light up a blast of gamma rays, and spew neutrons all over the nearby volume. Those rapidly moving neutrons slam into any nuclei in the way, processing heavier and heavier elements as they go. This mechanism is thought to provide about half of all the heavy elements enriching the universe today. So even though the kilonovae are less than a tenth as strong as their supernovae siblings, they pack just the right kind of punch to make an abundance of the heaviest elements in the universe.

Neutron star mergers are so violent that they're one of the few sources of strong gravitational waves around.[6] "Gravitational what?" Sigh. Remember that bit earlier about dancers and the dance floor? Oh, you slept through it? Well, skip back a few pages and refresh yourself, because that analogy was too good to use just once.

Imagine two dancers, stamping their feet to the music, probably off-beat because they have poor coordination. If they're heavy enough, not only do they make little dimples in the dance floor, but they can shake the floor itself, sending out waves. Like a waterbed. Dancers on a waterbed. OK, maybe that's not the best analogy. But it's the one you're gonna get, so you better be happy.

Because they are so incredibly dense, and at the moments before their ultimate collision they are orbiting around each other so quickly, they're able to vibrate spacetime itself. These ripples are almost unfathomably weak—gravity is already by far the weakest of the forces of nature, and these waves are tiny, effervescent perturbations on top of that—but they can be felt with sensitive enough detectors.

Note that these instruments typically involve bouncing lasers back and forth in mile-long tunnels with detectors sensitive to movements smaller than the width of an atomic nucleus, so it's not like you're going to carry them on your ship, but it's good to know if you're bored on the surface of a planet one day and need a project to kill the time.

But the violent weirdness doesn't end there.

It can get even worse. What's worse than a neutron star? Well how about a *quark star*?[7] Quarks? What in the cosmos are those? Oh, just the particles that make up protons and neutrons. Electrons are just electrons, but neutrons are a bag of three quarks and some glue. That's how atoms roll.

White dwarves are held up by the degeneracy pressure of electrons. Neutron stars are held up by the degeneracy pressure of neutrons. And quark stars—if they exist—would be held up by the degeneracy pressure of quarks themselves.

This is a really wild idea because quarks *hate* to be by themselves. Seriously, if you try to rip apart two quarks, you have to apply so much energy that new quarks spring in from the very vacuum of spacetime to fill in the gaps and buddy back up again. In addition, quarks are exceptionally picky about how they bundle up—there are only a few stable arrangements of quarks allowed in our universe.

But a quark star may have enough gravity to hold onto the quarks, keeping them together despite their inclination to drift apart. Just imagine that: a giant dead star, made of free-floating bits of one of the most fundamental building blocks of all matter in the universe.

We don't know if quark stars can even exist, and if they did, they wouldn't look much different from a neutron star. After all, once you've reached the density of nuclear material, there isn't a lot more you can squeeze in. So it could be that you encounter a neutron star that's nothing but a neutron star; or it could even be that the deepest cores of neutron stars are already a quark star in the making.

What we do know is that if quark stars are possible, they are not born quietly. The process that leads to a supernova builds a rapidly rotating, viciously magnetized neutron star. After a few million years, that neutron star could accrete enough matter and slow down enough that its innards start to transform from neutrons to free quarks. That transformation releases enough energy to make this *quark novae* a candidate for the most violent explosions in the known universe.

So one day you could be sipping tea, in orbit around a long-dormant neutron star. While you're orbiting, its insides have reached a critical threshold, igniting the exotic quantum processes that begin its transformation. In less than a heartbeat, the neutron star has transformed itself—and you're gone.

For now, though, they're purely hypothetical, so maybe—emphasis on *maybe*—you don't have to worry about them.

For now. It is a big and strange universe, after all.

✪

Stars are dangerous at their birth, throughout their lives, and even in death. It's a shame that they're so useful, providing heat and light and the right chemistry for life. If it weren't for that (some would say debatable) beneficial aspect, I would simply recommend avoiding them altogether. Stay where it's dark, it's safest there.

But no, we need energy, and stars are great sources of it. It's possible that habitable worlds could survive around white dwarves and, to a lesser extent, neutron stars. After all, their light output stays stable (kind of) for millions or billions of years. Assuming that any planets actually survive the death spasms of the original star, it may not be such a bad place to call home. There are certainly worse.

White dwarves and neutron stars are pretty much safe on their own, as long as you give them a healthy and respectful distance. But glitches, flares, and novae are enough to give even the most seasoned explorer pause. Is it really worth the trouble to enter this seemingly dead system? To explore and try to understand the complex physics dominating the forces there? Or will that dead star rise from the grave, even briefly, to drag you down into the underworld with it?

Could there be anything worse than a neutron star? Something even more dense?

Yes, there could be. I'm glad you asked.

The birth of a neutron star is perhaps the most violent single event the universe can ever produce. Overwhelming gravity, exotic quantum forces, floods of rare particles, and good old-fashioned turbulence all conspire to turn a star inside out in less than a second.

The absolute most important ingredient, the piece that sets the whole thing in motion to its violent end, is the formation of that protoneutron star in the core. It's the resistance offered by that bizarre ball of neutrons that triggers the explosion. And in that moment of fury, at the onset of the passionate wrath that will engulf the star, its own collapsing atmosphere—tens or even dozens of suns' worth of material—must rebound off that core, transforming its gravitational energy into raw power.

Sometimes that core survives the moment of impact and continues on in life as a neutron star.

And sometimes it's too much.

Just as the degeneracy pressure of electrons can be overwhelmed, forcing white dwarves to collapse in on themselves, so too can the degeneracy pressure of neutrons, and even quarks. There is a limit to what they can sustain.

But once a neutron star can't support itself against irresistible, insatiable gravity, what then? What last redoubt is available to the core? What bastion of safety and support can it turn to against the onslaught of gravity gone berserk?

Nothing. Nothing can.

In those fiery moments that signal the end of the life of a proud, massive, blazing star in our universe, when the conditions are just right, when the weight of the crumpling atmosphere is just too much, something far stranger, far more exotic than a neutron star merges from the wreckage.

Supernova explosions don't just make neutron stars. They are the forges of the black holes.

Supermassive Black Holes

They say you can still see him
If you look carefully
Waving goodbye
Or waving for help.

—Rhyme of the Ancient Astronomer

P oor Alice, we will remember her sacrifice. Too bad about the whole getting stretched like a piece of taffy out into a thin stream of atoms. But, what a way to go! And we got to learn a little bit about the dangers of small black holes. Bob, meanwhile, has already moved on and found a new partner in space exploration: Carol. Carol is much smarter than Bob, as we will see when we follow them to their next destination.

Bob and Carol travel to someplace plenty of young travelers—people just like yourself—often want to go. The Grand Central Party. The Big Bulge. The Bottom of the Well.

The Galactic Core.

They go, lured by the lights, the sounds, the action. Billions of stars, crammed into a central bulge, hustling and bustling in their lives. Complicated physics! Radiation! Magnetic fields! Nothing like the sleepy, dopey suburbs of the galactic disk and our home system.

Do you know the origin of the term "siren's song?" It's a Greek myth. This group of creatures known as the sirens would hang out near some rocks, singly sweetly to any passing sailors. But it was more than just a pleasant melody: their tones would make them irresistible to the sailors, who would be seduced beyond reason, lured by the music. The sailors, spellbound, would steer their ship to attempt to get as close as possible to the sirens, crashing onto the rocks.

And then the sirens would eat the sailors.

And I'm telling you now: there's a menacing beast in the heart of our galaxy, ready to eat you if you get too close. A dark heart in an otherwise brilliant galaxy. It will sing to you, sweetly. You can hear its song from light-years away. Sail too close and you'll feel the compulsion—the need—to get closer and explore further. Who knows what you might find, deep in the heart? What new physics awaits you? What brilliant panoply of fantastic forces and particles?

You'll want to party at the core, spending a lifetime exploring its mysteries,[1] but this lurking nightmare will consume you in a flash, whole, alive, picking your bones apart before you've even had a chance to notice that it's too late.

This creature is nothing like those puny stellar mass black holes that roam in the interstellar wastelands. Those are ants, flies compared to this beast. No, this is something far larger, far older. It has been feeding for millions—even billions—of years. An ancient terror. Almost as old as the universe itself. It was here when our Milky Way galaxy first formed, and it will remain even after all the billion suns have been extinguished.

Our entire galaxy is anchored on this black horror. It's quiet now, slumbering for at least a hundred million years. In the past, when it was awake, the galaxy quaked and trembled in bouts of fire and rage, a tempest so fierce that stars themselves stopped forming. For now, we are safe from its wrath, but should this dragon be provoked, it will spit its fire for a million light-years.

There, buried in the deepest part of our galaxy, surrounded by retinues of small black holes and a host of giant stars, is the beast itself. We humans are not meant to be there, not made for such extremes. It's there, in the heart, where a massive creature of pure gravity and malevolence sits and hates, reminding us of our cosmic insignificance.

I almost dare not speak its name, but I must, to warn you of what you might encounter should you travel inward.

I speak of gravity. I speak of blackness. I speak of infinities. I speak of the unknown. I speak of the void.

The dead center at the dead center of the galaxy.

I speak of Sagittarius A*, a supermassive black hole.

✪

Earth astronomers gave a name to the beast a long time ago. It was a surprise—from their vantage point, there's nothing to indicate anything particularly worrying about the constellation Sagittarius. We had known for decades that this was the direction of our galactic center, the middle of the Milky Way, deduced from careful observations of the movements of globular clusters—small clumps of red and dead and nonthreatening stars—orbiting our galaxy, if you just look naively with a telescope, we're surrounded by stars on roughly all sides, but globular clusters obviously *don't* orbit around us, betraying the true heart of our galaxy.

Still, besides just being denser than average, we never suspected. What tipped off those early observers was an intense amount of radio energy emanating from a specific location within that constellation. The brightest deep-space source of radio emission in the universe. It was just one more mystery to add to the pile, and so the innocuous sounding name of Sagittarius A* was assigned and life went on.[2]

They had no clue what they had just revealed. And they're probably thankful that they got to live their lives in peace and quiet, safe and secure on our home planet, observing and revealing the nature of that giant black hole, knowing that they'll never have to get too close, and that the chances of a vengeful awakening are blissfully minuscule.

But still, observation after observation exposed the true nature of this strange but strong radio source. They found the black heart of the Milky Way.

But don't think you can avoid supermassive black holes by moving to another galaxy. Oh no, every person you know has a dark secret, and every galaxy we see has a black heart. Most of the raw material in a galaxy— the gas, the stars, even the dark matter (we'll get to that later)—sits in the center, because that's the way gravity likes it. So you've got a lot of stuff, with lots of gravity, and nothing to stop it, and before you know it you've formed a black hole.

Galaxies form slowly, starting off small in the early universe, gradually gobbling neighbor after neighbor. The biggest galaxies are formed from countless mergers and acquisitions. And the central black holes, the dark secrets, started off small, too. But over the eons gas rains into the core, finding its way through the event horizon and making the black hole bigger.[3]

But that's not the only way the black holes grow.

Big, mature galaxies like the Milky Way form from the mergers of smaller galaxies, and those galaxies are already hosting black holes (in fact, we see evidence for giant black holes basically just as soon as stars and galaxies alight in the early universe, which is something of a problem in astrophysics, but that was a long time ago in the history of the cosmos so it's nothing we need to worry about right now).

When galaxies merge, where do their big black holes go? Gravity does all the work we need it to do—the black holes find each other and start to dance. But it's unlikely that they would hit directly head-on and merge right away; the odds of that kind of lucky collision are just too low. Instead, the two black holes come close enough to get mutually gravitationally captured and orbit each other.

And here's where we have a bit of a mystery. Black holes orbiting each other is fine, who cares, no biggie. And orbits are usually pretty stable: The Earth isn't going to crash into the sun anytime soon, and two orbiting black holes probably won't, either.

But we don't see orbiting black holes.

We just find single, solitary, supermassive ones. So how do they merge?

One way is through gravitational waves.

Our two orbiting black holes can send off gravitational waves—ripples in the dance floor of our universe, in the very fabric of spacetime. This takes energy (it's not easy), and when they lose energy, they get a little closer together. Problem solved? Not quite. The issue is that this process of spiraling in toward each other by the emission of gravitational waves is really slow, like lifetime-of-the-universe slow. And since it's pretty obvious that black holes have merged in less time than the lifetime of the universe, we're a bit stuck.

Earth astronomers have dubbed this "the final parsec problem."[4] They run fancy-pants computer simulations and can watch as big black holes get to about a parsec (about 4 light-years. I challenge you to work the word in casually in a conversation at a cocktail party. For example: "Did you see how sweaty Ted is? It looks like he ran a parsec to get here!") of each other, but no closer. There are some tricks they've invoked to make the problem go away, but nothing they're too sure about.

Hey, the universe still gets to have *some* mysteries, right?

★

However they manage to do it, black holes do manage to merge. One after another, collision after collision, the black holes grow. And when they're not merging, they're feeding, suckling on the rich supply of matter in the centers of galaxies. Atom by solitary atom, they gorge to become not just massive, but supermassive.

And when I say *supermassive*, I'm not just being cute. We're talking at least a million times more massive than the sun, and that's just the start. We're not playing games anymore, kid. The largest black holes in the universe easily tip the scales at hundreds of billions' worth of solar masses.

With those kinds of masses, these black holes are truly big in every sense of the word. Still minuscule compared to a galaxy—while our own supermassive black hole is the single most massive object in the Milky Way, it's still less than a fraction of a percent of our total mass. And as far as size? Sure there are certainly larger things out there, but they're all extremely tenuous, like Oort clouds and nebulae.

Placed in our own solar system, Sagittarius A*, with a mass of 4 million suns, would stretch its event horizon out so far that it would engulf the sun and reach halfway to Mercury. The largest black holes we've found could swallow the entire solar system whole, without even blinking.

You'll also note that I've only talked about two sizes of black holes: "small and shrimpy" and "big gulpy." It seems like there are only two varieties of black holes: the smaller ones ranging anywhere from a few times the mass of the sun to fifty times bigger, and the not-just-massive-they're-supermassive ones starting at *millions* of times the mass of the sun.

What gives? Shouldn't there be medium ones too? Where are all the black holes a thousand or 10,000 times bigger than the sun? Why the big divide? Surely the supermassive ones had to be medium-massive at some point in their journey to extreme corpulence, but as far as we can tell the process of black hole gigantism happens relatively quickly, before anyone in the universe has a chance to notice. And once settled and evolved, the huge black holes simply remained in their lairs, undisturbed. Since all this megafeeding and supermerging probably happened pretty early in the universe, when galaxies first started getting their act together, we don't expect to see any of the medium ones remaining: they're all fat now.[5]

The small black holes remain small because they're isolated and lonely throughout the immense volumes of their home galaxies. But why don't the growing black holes simply stop halfway to supermassive? Is the universe simply incapable of making black holes of that size?

Maybe, maybe not.

Galaxies like the Milky Way host big black holes, but maybe smaller galaxies host . . . smaller black holes. Maybe if we looked in really small dwarf galaxies, they might host black holes that never got a chance to grow to greatness. Earth-based astronomers have found some slim evidence for their existence, but if they do exist, they're a) very rare, and b) don't give off a lot of the bright radio emission that we see from something like Sagittarius A*. And I'm perfectly aware of the fact that you might be wondering how a giant black hole could possibly be the source of copious amounts of radio waves, but the resolution to that particular mystery is going to have to wait for its own chapter. Don't worry, it will be worth the wait.

It's also possible that the central core of every galaxy hosts innumerable medium black holes (which are called, of course, *intermediate mass black holes*), that are largely undetectable because, like their smaller cousins further out in the galactic disk, there's nothing falling into them and hence nothing to light up a warning sign around them. These medium black holes may be in the process of slinking down further toward the center, caught in the act of seeking the final embrace of the giant in the center.

I've got a solution. Here's a homework assignment: go to the nearest dwarf galaxy and let us know if you see any mediumish black holes. Don't try going to the galactic center, because there are just far too many dangers in that direction. But dwarf galaxies are about as benign as they come; much safer to seek your answers there. Or just give the astronomers another decade to sort things out. Your call.

The bigger the party the bigger the danger: larger galaxies host more massive central black holes. Supermassive black holes in the hearts of galaxies appears to be mandatory. We see them *everywhere*, in *every* galaxy.

✪

Finding the black hole in the center of our own galaxy is easy. Easy for astronomers, at least, because of the insane amount of radiation and activity nearby, a scaled-up version of what goes on around small black holes in binary systems. But all the heat and light only tell us about the gas and dust swirling in toward the ever-open maw of the giant black hole, not about the beast itself.

But some are foolhardy enough to venture too close, and from their mistake we can glean useful knowledge.

Around Sagittarius A* we see lights. Stars. Orbiting. Dancing. Daredevils. Thrill seekers.

Fools, all of them. An entourage of stars, locked in mad orbits around the black hole itself.

I'll say it again to make sure this fact sinks in: We can directly observe, even from the vantage point of the Earth over 25,000 light-years away, the motion of stars around the black hole in the center of the Milky Way galaxy. Keep reading that sentence until your mind is sufficiently blown. As those stars whip around, sometimes reaching over 2.5 percent the speed of light as they career close—far too close for comfort—to the event horizon, we can measure their speeds and identify the center of their orbit, which couldn't be clearer: nothing at all.[6]

And from calculations of gravity we can estimate the mass of the central monster.

While this technique allows us to take the measure of Sagittarius A* without ever having to visit it (thankfully), this trick doesn't work for faraway galaxies, where we can't observe the individual stars in orbit around their black holes. But we're not hopeless, and Earth astronomers are abundantly clever, and we can measure the motion of gas in the center of a galaxy, and through that get a better understanding of the supermassive black hole, with the Doppler effect.

You know, Doppler effect? OK, I'll explain it. But just once, so you better pay attention.

Let's say you're standing far away from someone, and the two of you are holding opposite ends of a giant slinky. Why? This is an analogy, just roll with it, OK? If your friend starts running toward you, the slinky will get squished up, and after she passes you, it will stretch out again.

This works with sound: if she was screaming (probably saying "Why are we doing this?"), as she runs toward you the sound waves will be squished up into higher pitches. And opposite as she runs away.

Here's a visual demonstration: "ooooooooaaaaaaAAAAAEEEEEEE-AAAAAAaaaaaaoooooo."

You hear this all the time, like when trucks or ambulances drive past you: not only does the volume change, but the *pitch* changes as well. But you were probably too busy asking yourself what you're doing on the side of the freeway.

It works on sound and it works on light. If a ball of gas is spinning as it falls into a black hole—as it often does—one edge will be spinning toward us and the opposite edge will be spinning away from us. So the light from one side will be squished to higher frequencies (toward blue, hence *blueshifting*) and the other side will be stretched to lower frequencies (toward red, hence *redshifting*). So by making careful observations (note: astronomers are never fond of careless observations), we can figure out the mass and size of the central object and identify it as a black hole.

Hey, it ain't easy, but there's a reason "astronomer" isn't a synonym for "lazy."

And if that's not enough, we can also get pictures of them. Literal, living, breathing photographs of black holes.[7] Now you might be wondering—rightly so—how we can take a picture of a black hole, which is as black as the space behind it. It's the same problem as with their smaller cousins, and just being big isn't going to make it any easier.

As (super)massive as these black holes are, they are also ridiculously far away from any planet-based astronomer. Sagittarius A* itself is 25,000 light-years away, and the monsters in the hearts of other galaxies are millions or billions of light-years. Even the biggest beasts are the tiniest dots on the sky, impossible to resolve unless we had a telescope the size of the Earth itself.

So, astronomers built a telescope the size of the Earth itself. Like I said, clever folks.

But it wasn't a single massive dish, as awesome as that would be. Instead it was a network of telescopes, scattered across the globe, all trained on a single target simultaneously. This is in effect a virtual telescope as wide as the Earth, giving it enough resolving power to capture images of distant black holes. This kind of astronomy—known to the nerds amongst us as *very-long-baseline interferometry*—isn't easy. The telescopes, data needs to be paired with precise atomic clock outputs so everything can be synchronized, and they all need to stare precisely at the same spot on the sky for days or even weeks at a time.

The result from this Event Horizon Telescope was exactly as advertised: an image of an event horizon. The first was the massive but relatively calm beast in the center of the galaxy M87, weighing in at over 6 billion suns and a radius that would stretch to the inner edge of the Oort cloud.

But, right, event horizons are black. Thankfully the Event Horizon Telescope didn't try to take a picture of the event horizon itself, but the stuff around it: the disk of accreting material swirling and screaming its way into oblivion. And because of the extreme bending of spacetime around the black hole, light from the accretion disk *behind* the event horizon curves up and around the thing itself, letting us see around it.

The end result: a ring of intense radio light, haloing a center of pure blackness, of nothingness, a void. A silhouette of a black hole.

Supermassive black holes appear to be a fundamental part of the universe, but you visiting them is entirely optional. You know how Mother Nature likes to warn you about dangers, like a green frog is saying, "Go ahead, lick me, it's only gross" and an orange frog with purple spots is saying, "Go ahead, make my day?" The central black holes are covered in clouds of hot, bright gas, emitting lethal levels of X-ray and gamma ray radiation, enabling us to see them from across the universe.

Mother Nature is abundantly clear here: please, don't lick a supermassive black hole.

✪

Now for the juicy bits. Bob tries to convince Carol to explore the central core of our galaxy. Besides the deadly radiation, relentless cosmic rays, and frequent explosion from the death of a massive star, it's a pretty cool place.

But Bob wants to get closer, to tease the dragon itself.

Carol, wiser, decides to stay back, and that choice will allow her to get a very different picture of what's about to happen. That's because there's a reason that the word *relative* appears in general relativity. Observers in the universe never agree about the length of objects and the duration of events. If you have a different frame of reference—a different speed and direction—you will have a different view of the universe around us. It's baked into the fundamental structure of spacetime (indeed, it's what allows *spacetime itself* to be a thing).

Different observers always disagree, and black holes are the ultimate divorce.

Here's Bob's perspective.

From a distance, the black hole is quiet and calm. Our own, Sagittarius A*, hasn't been on a feeding frenzy in a while and is dozing quietly. There is a small swirl of accretion disk around it; nothing that tremendous given what the black hole is truly capable of. By estimating the mass of Sagittarius A*, Bob can calculate the event horizon and figure out a safe distance to watch from.

He decides to get closer.

As he approaches, the event horizon swells beneath him, appearing far larger than it should, given its size and distance. The geometry you learned in high school simply doesn't apply here; the extreme gravitational distortions allow you to see around and behind the black hole, making the event horizon loom larger—and darker—than life. The accretion disk too seems to wrap the black hole in a ring of blinding light.

The descent continues.

As Bob gets closer, the gravity ratchets up in intensity, pulling him more and more strongly. He can turn around and escape if he chooses, but it's getting harder with every passing second. One way to think about the approach to a black hole—one that's perfectly allowed by the equations of general relativity—is to imagine a black hole as a sinkhole, a place where space itself flows inward like a river flowing toward a waterfall.

As long as Bob stays outside the horizon, he can turn around and battle his way upstream, but the event horizon is the edge of that infernal waterfall, where the water—space itself in our analogy—is flowing inward at the speed of light. After passing that, Bob can try to turn around and fight it all he wants, but since he can't ever go faster than the speed of light, he'll continue to fall ever inward, doomed.

The descent continues.

As Bob nears the event horizon, his speed approaches that of light. Or—and this is especially frightening to think about—yet another viewpoint is that the event horizon begins to approach *him* at the speed of light, and he will have to work harder and harder to avoid its onrush.

Bob's not going insane: things like the exact locations of event horizons can only be pinpointed from a distance. Up close, the boundary is not easy to find, and appears to be moving toward you the closer you get. That's because you're getting closer and closer to The Limit—the point of no return, where you would have to escape only by going at the speed of light. In other words, you have to *outrun* the event horizon by speeding away from it. Get too close, and you'll never be able to leave faster than it's chasing you: you're caught. Spooky stuff, but they're not called "smiley happy holes" for a reason.

The descent continues.

What will Bob encounter at the event horizon? Some theorists believe, for various arcane reasons, in something known as the *firewall*, a scorching hot inferno of fundamental particles, born out of the vacuum of spacetime itself, as energies high enough to immediately incinerate anything falling inward. This firewall is invisible to outside observers, since it's right at the event horizon itself, light from this boundary cannot escape, it can only be experienced.[8]

But we're not sure if Bob will encounter the firewall and, unfortunately for us, he'll never be able to send reports of his survival.

According to general relativity, however, what Bob sees at the moment of crossing the event horizon is . . . nothing special. In fact, there's nothing to signify the rite of his passage through that boundary. There's no light. There's no warning. There's nothing special about the universe around him. He simply . . . falls.

Before Bob knows it, he's gone too far. Literally, he may not be able to gauge the exact moment he slipped over the line and through the event horizon. To Bob, it was just like any other moment. The event horizon isn't like a wall between two rooms. It's more like a wall between two *futures*.

The descent continues.

As Bob crosses into the black hole, he can still look out: light from the rest of the universe falling with him and after him strike his retina, allowing him to witness the universe that he will never, ever again be able to join.

And yet, something strange has happened to Bob's view. Even though his calculations tell him that he has finally crossed the event horizon, it still appears below, swelling larger and larger, as if he were still approaching it. He has indeed crossed into the void—his calculations were faithful to him—but he's experiencing one of the largest tricks of perception that black holes can play: the false horizon.

✪

To understand what Bob is seeing, we need another quick relativity lesson. I know this is getting rough, but hang in there. You need to understand this in order to develop a healthy respect for black holes in the wild. Trust me.

Here's a fact of nature: moving clocks tick slow.

Moving clocks tick slow. Moving biological processes tick slow. Moving *Bobs* tick slow. Moving *yous* tick slow.

Take two ultra-super-high-precision atomic clocks. Leave one just sitting around, like your brother-in-law. Put the other one in a jet and whiz it around the world for a few hours. The jet-setting atomic clock, when it gets back from its worldwide tour, will be a few fractions of a second behind the stationary one.

Sign, sealed, delivered: this is the way the world works. I know you don't believe me (why should you), but that's not important. What's important is what happens to Bob.

So what does that have to do with Bob? He's not moving, right?

Here's where our friend Albert comes back. The guiding principle behind general relativity, the one postulate that forms the cornerstone for entire field of physics, is that there is no difference between you feeling the accelerated push of a rocket and you feeling the gravitational pull of the Earth.

I know, I know. Have a seat, take a breath. A sip of water. We'll get through this together, because it's important. Bob's about to breathe his last, and we want to understand his final moments. Don't do it for me, do it for Bob.

You, reading this, feel the gravitational pull of a planet, probably the Earth. It constantly accelerates you downward with a certain force—1g, conveniently enough. Now let's say we kidnapped you, put you inside a rocket ship, closed all the screen doors and windows, and blasted you off. And let's say we gave that rocket a swift kick in the rump and made it accelerate at a constant 1g. (Note: physicists make *excellent* practical jokers.)

Could you tell the difference? Without looking out the windows, could you perform any experiment or make any observation to tell if you were on the earth versus on a rocket ship?

Einstein said *nein*, you can't. And that simple fact gives rise to the entire theory of general relativity. Oh, and lots of math, too, but you get the idea.

Back to clocks. Moving clocks tick slow. Accelerated clocks tick slow. Gravitationally pulled clocks tick slow.

It's starting to click now, isn't it? I can hear your brain all the way from here.

A clock on the surface of the Earth will run slower than a clock far away, because of this equivalence principle. Put a clock way high up for a while, then bring it back down. The Earthbound clock will be slower.

"I give up," you say, "relativity is freaking me out." It's OK, we've all been there. Einstein may have been a genius, but he didn't have to be right, after all.

Except, he kinda was. We tested it. In fact, you might be testing it right now. Ever use GPS? Grandpappy's pancake syrup? It's the best. It also stands for global positioning system; it's the set of satellites that tell you how lost you are and how late you are for work.

If the GPS system didn't correct for this clocks-running-slower effect, they would be *wrong*. So the next time your smartphone correctly tells you your location, you can pretend you're a scientist and that you just performed a test of general relativity. Congratulations.

The stronger the gravity, the more equivalent it is to a harder acceleration, and the slower the clock. So while you may only be a few microseconds behind an orbiting astronaut, Bob is *way* behind.

✪

Finally, back to the black hole, and an understanding of Bob's false horizon and what Carol, sitting safely outside, witnesses of the event of Bob falling to his doom. Ah Carol. Wise, wary Carol. A good role model. Maybe you should read her book instead. She views Bob's descent from a safe distance. And she tells a completely different story: from Carol's point of view, Bob never even enters the event horizon.

"Flabbergast!" you might say. After all, I just wasted a lot of your time and mine describing Bob's experiences inside the black hole. What gives?

Relativity gives, that's what.

Bob's at the bottom of a well—a deep well in spacetime created by the mass of the black hole. As he approaches the event horizon, he goes further down, down, down that well. Carol, up top, sees Bob getting dimmer and dimmer.

But also redder and redder. The light emitted by Bob (Maybe he's carrying a flashlight to see inside the black hole. Get it? It's a joke.) has to climb all the way out of that well. And that takes energy. Light that loses energy gets stretched to longer wavelengths, down from blue . . . to red . . . to infrared . . . to microwaves . . . to radio. The closer Bob gets to the event horizon, the redder the light that Carol sees.

I'm not joking; We've tested this. Shine a laser up to the top of a very tall building. Record the wavelength. Redder. I guarantee it.[9]

Because Bob's at the bottom of that well of gravity, he's also slowing down. Bob doesn't notice anything different to himself. He's still Bob, but our view of him from the outside is what changes.

So that's what Carol sees: a slower, dimmer, redder Bob. A Bob that never *quite* reaches the event horizon. From her perspective, far way, Bob takes an infinite amount of time to cross the threshold. Sure, every time she checks, Bob has gotten fractionally closer, but she'll never witness the crossing itself. That's because the light that Bob emits *just* as he enters the event horizon is already too far gone: it would take an infinite amount of time for that light to reach Carol. I don't think that she's willing to wait that long.

It seems weird that Bob—according to Bob—can cross the event horizon, but Bob—according to Carol—never quite does. I told you, black holes are no light stuff.

This is one of the points of relativity. I've said it before and I'll say it again: the perception of events, the measuring of time and distance, are all *relative* to the observer.

And Bob's false horizon after he passed through the event horizon is the ultimate price of relativity.

You see, what we see from the outside of a black hole isn't the event horizon. The event horizon isn't a *thing*, it's simply a mathematical marker that denotes the line where space is rushing inward faster than the speed of light. It's not a physical object.

The blackness at that boundary is more of a . . . memory.[10]

Black holes are born from the deaths of massive stars. As those stars collapse and collapse, their plasma bodies cross that line and keep going. But us external observers, watching from a safe distance, never actually get to witness the formation of the black hole—the last bit of light from the dying star, like the last bit of light from the dying Bob, is frozen at the event horizon.

And so when Bob passes through, there's nothing to mark or signify his passage. The pitch blackness of the event horizon was a lie, and the true blackness still sits below him.

What he sees, what appears to grow larger with every passing second, is the true heart of the black hole. The true center. The true death.

The singularity.

★

Right, the singularity. There it is. I finally said it. I've been trying to avoid it this whole time, but now there's no way around it. There's a joke there, if you've been paying attention.

All the stuff that gets sucked into black holes—gas, particles, people—ends up compressed into an infinitely dense point. That's the result of overwhelming gravity that cannot be defeated. That point is infinitely tiny, too: a literal geometric point, having no spatial extent whatsoever. It's hard to wrap a human-sized brain around the concept, and it seems wacky and unnatural, but it's predicted by general relativity. Now we know that our current understanding of physics gets a little bit fuzzy at very high densities and very small volumes, but for now let's just run with it, because it's not like anybody's gone into a black hole and reported back what they've found.

Back to Bob. One he's passed the event horizon, he will, no matter how hard he fights it, end up at the singularity. That's the point of the event horizon: it separates his future possibilities.

Outside, he can have any path in any direction he wants; inside, the singularity is like Rome: all roads lead there.

Another way to put it: outside a black hole, you can go anywhere you want. You can explore distant galaxies. You can fulfill your dreams of studying toe fungus. Whatever; the universe is your oyster. But you can't avoid the passage of time—you can't avoid your future. Wherever you go in space, you must always go forward in time.

But inside the black hole, the singularity *becomes* your future. All your futures. This was that the nightmare that Schwarzschild found at the event horizon and made him recoil in terror. It simply *didn't make sense*.

But black holes are all too real.

No matter where Bob looks, he sees the singularity. No matter what direction Bob moves, the singularity stays in front of him. It's a single geometric point, but because of the extreme gravity—the most extreme gravity in the entire universe—it appears to swell and grow. A hole in the universe, carved out from spacetime itself.

The black singularity expands, filling and flattening as it does, appearing as a vast featureless plain. As if landing on a giant black Earth, Bob nears his ultimate oblivion.

At the moment the blackness appears to stretch perfectly from horizon to horizon, cutting out half of his sky, Bob will have reached the singularity.

How long does Bob have before his toes touch that infinitely dense point? It depends on the exact size of the black hole, but for supermassive black holes, it's only a few minutes.

That's it: a few minutes. And there's nothing he can do about it. The singularity, remember, is in all his futures. Bob can't avoid the singularity the same way that you can't avoid tomorrow. It is his final end. His best option is to simply ride it out, letting himself fall freely. If he tries to fight it, that will only hasten his rendezvous with annihilation.

I told you, we were not meant to be here.

Oh, what about those tidal forces that did fascinating things to Alice on her approach to a stellar-mass black hole? Why does Bob get to make it through the event horizon in one piece? There are a few pieces to this puzzle, so here's a list: 1) all the mass—and I mean ALL the mass—of a black hole is concentrated in the infinitely dense, infinitely tiny singularity, 2) the more massive the black hole, the bigger its event horizon, and 3) all those tidal effects depend on how far away you are from the massive object.

Put all this together—go ahead, take your time—and what do you get? For supermassive black holes, the event horizon is so far away from the singularity that the tidal effects are incredibly small at that boundary. Bob will meet his pasta-like fate for sure, but only *after* he's passed through the horizon and he's closer to the singularity itself.

Note that this only applies to something relatively small and compact like Bob. For something bigger, like an unlucky star, it can be ripped apart by the extreme gravity. These events, known amongst polite society as *tidal disruption events* and to more uncivilized folks as *a star getting torn to shreds*, are some of the most energetic incidents in the universe, capable of flaring so brightly that they can be seen from billions of light-years away.[11] Suffice it to say you should stay well clear from these death screams of a hapless star as its falls into the final, deadly embrace of a giant black hole.

But Bob? Small enough to slip through unnoticed, the great eye of the black hole focused elsewhere.

Bob gets one more treat before he becomes a treat for the singularity: he gets to watch some of the history of the universe unfold before his eyes.

Assuming, of course, that his eyes remain intact. But he only gets to live for another few minutes (plus or minus a minute, as if that counts), so how does that work?

It works because Bob is slow. You're slow too, and I'm not making fun of you. Bob is slow because he's near a black hole, and you're slow because you're probably sitting on the surface of a planet.

To be fair, both you and Bob don't *feel* slow, do you? It's only relative to someone else, say, Carol, who's far away, that you seem to slow down. And there it is, our favorite word these past two chapters: *relative*.

Generally relative, that is.

To Bob and to you, nothing seems out of the ordinary. It's only when a distant observer tries to compare clocks with you that things seem off. So to Bob, he only gets a handful of minutes, but from his perspective the entire universe has sped up to be about twice as fast.

He still gets to see the light from the outside universe—heavily distorted and blueshifted because, hey, black hole, plus squeezed down into a donut-like ring around his waist, because of the extreme gravity—but still watchable.

And then Bob hits the singularity and becomes one with the infinite density. What happens then? Who knows, some problems are hard even for me. Why don't you jump in and find out yourself?

Suffice it to say, Bob's last few moments of existence and awareness in this universe will be . . . interesting.

✪

OK, so maybe I lied. Maybe not. Truth is, we don't know for sure what happens inside an event horizon. Outside the black hole, we're in good hands: we can do tests and experiments and other sciencey stuff to validate Einstein's legacy. But inside? Well, technically, we can't access it. Which means we can't make observations. Which means we can't make tests. Which means we don't know *for sure* about stuff like the singularity.

But we do have math, and math is powerful. The theory of general relativity has been subjected to a plethora, a variety, and smorgasbord of tests. I mean, who *wouldn't* want to be the one to finally prove Einstein wrong? That's Nobel territory right there. Come on down and collect your free trip to Stockholm.

But general relativity keeps on truckin'. And it makes very precise predictions for what happens inside a black hole. Now at the very tiniest scales we know it has to break down—quantum mechanics intervenes at some point, and like I said, we're kinda hazy about some of the details—but above that scale we're in familiar territory. Until he hits the singularity, the gravity that Bob would experience inside a black hole's event horizon is no different than what you and I are experiencing *right now*, sitting comfortably on a planet's surface. Oh, yeah, except a lot stronger.

So jump right in, pal, the water's warm.

No, wait, don't. It's a bad idea. Assuming you can survive the deadly radiation surrounding a galaxy's central supermassive black hole, assuming you can survive the descent through the event horizon, assuming you can figure out a way to maintain your bodily integrity as you spiral in to the singularity, you only get a few minutes of fun before it's lights out.

If you hear some huckster trying to sell you some fancy gizmo for navigating black holes, ignore them. I can't tell you how many foolhardy explorers thought they had some brilliant scheme for escaping an event horizon.

We're still waiting for them to report back.

I should note that it's possible that black holes don't exist. I know, I know, then why did I blather on for so many words about it? Better safe than sorry, that's what my mother used to always say, before getting too close to an active quasar. In her defense, it didn't seem active at the time. Earth-based astronomers assign a very high likelihood to their existence, especially considering we have pictures of them and everything.

When Schwarzschild first figured out all the monkey business about event horizons, most physicists just assumed it was a fluke. Sure, the equations of general relativity *allow* for these weird "black holes," but that doesn't mean that Mother Nature has to actually provide them. After all, lots of things are developed in mathematics that don't appear in nature. It was thought that nature would always find a way to beat the Final Collapse and avoid the formation of a true black hole. Maybe by having a last-minute reaction that blows apart the collapsing star.

Maybe by having additional physics that support a dead star again the crush of gravity.

Event horizons certainly appear to exist and do everything that event horizons are supposed to do according to Einstein and Schwarzschild. But the singularity? That's a different story. We *know* that singularities don't really exist, that nature does replace them with something else. But we don't know what, exactly, and we may never find out.

But these monsters look like black holes, act like black holes, and talk like black holes. The term *black hole*, with all the panic-inducing physics that go along with it, is the best and simplest explanation we have for those Earth-based observations.

So, until further notice, let's act as if black holes exist. And let's be terrified of them.

Quasars and Blazars

A profound voice
A booming sound
Echoing across the cosmos
Can you hear it?
　　　—Rhyme of the Ancient Astronomer

You're at a restaurant, and it's been a great night. Good friends, great stories, hearty laughs. Appetizers, salad, a little bread, the main course. Something fluid to wash it all down. Ready for dessert? You're not sure. You can feel you're near your limit. The pressure in your belly is almost unbearable, but you're having such a fantastic time that you don't want to stop. And that cheesecake looks so . . . *tempting*. Maybe just a small bite. Want to split a slice? You're pretty sure there's some room in your esophagus.

You get full after eating too much. But you're not a black hole, are you? Black holes are never satisfied. Their bellies stretch all the way down to that infinitely small, infinitely dense singularity. A black hole could swallow every atom in the universe and still be licking its lips, ready for more.

That insatiable hunger, that never-ending need, is an engine. An immense machine operating the most powerful instruments in the universe, lighting up beacons that stretch from one end of the cosmos to the other. Lighthouses with cycles lasting millions of years, carrying enough fiery intensity to heat entire clusters of galaxies.

And just like lighthouses, these cosmic flares serve two purposes: to offer help and guidance, and to warn of dangers.

Every galaxy hosts a supermassive black hole; we've been down this road before. And the bigger the galaxy, the bigger the black hole. The biggest galaxies are found near the centers of galaxy clusters (which, in case you've never met a galaxy cluster before, is a single collection of a thousand galaxies or more, bound together in an endless waltz of gravity), where all the gas finally finds a resting place. That gas is entombed in truly super supermassive black holes: hulking monstrosities well over a billion times more massive than the sun.

They didn't get to that size by being shy about what they ate. Gas, dust, stars, whole cows. It all goes down the hole. When the central black hole is in an especially voracious mood, when it's horking back shovelfuls of gas, its host galaxy gets a new name: it's called *active*.

What an innocent-sounding term, *active*. Like it's just playful or frisky. Sure, yeah, let's play make-believe.

Those giant black holes are appropriately fearsome, but you don't have to even approach the outer limbs of their event horizons before you reach the limits of Safe Space. The cores of these active galaxies are simply . . . something else.

That activity can be seen from the furthest reaches of the universe; you can imagine just how terrible it must be to be near one.

Despite their incredible brightness, however, it took centuries for Earth astronomers to first spot them, and they didn't do it through optical or infrared astronomy—it wasn't until they turned on their radios that they heard them. Besides Sagittarius A*, astronomers in the middle of the 20th century started finding interesting little points of intense radio light, which was understandably confusing.[1]

At first their intense brightness convinced the astronomers that they were seeing a new kind of object scattered within our own galaxy. But then they saw more, and more, and more, dotted all across the sky. Whatever these things were, they were definitely extragalactic. But to be that bright in the radio spectrum *and* that far away meant they had to be . . . so powerful they were almost unholy.

With their crude instruments, they couldn't see any structure to them—they were just dots, a lot like distant stars are just dots. So these things kind of looked like stars, but definitely weren't. Objects that were quasistellar. Quasistellar objects. *Quasars.*

This is a typical astronomy professional way of saying, "we have no idea what we're looking at," which at the time of their discovery, was true.

Further observations found some quasars to be extremely, uncommonly, inhumanely bright, and were named *blazars*, which is apparently short for "blazey quasar," or at least it should be.

As technology advanced, as it usually does, Earth astronomers began to get better pictures of these strangely intense objects, and found that quasars and their brighter cousins the blazars were both were just special kinds of what astronomers began to call *active galactic nuclei*, or if you're too busy in life, *AGN*. "Active" since they're not boring; "galactic" since they're a part of a galaxy (no points awarded for figuring that one out); "nuclei" because they happen in the rich creamy nougat center of those galaxies.

We'll get back to labels later, because classification is important as a safety measure, but for the purposes of your elucidation I'm going to stick to the word "quasar," because in my good and fair judgement that is by far the most awesome word to use when describing them.

✪

First, an anatomy lesson. We'll start our journey in a cluster of galaxies, a bustling metropolis of a thousand individual galaxies or more, plus a gigantic but thin atmosphere of hot gas. (They're the largest gravitationally bound structures in the universe, so they're worth remembering.) Zooming in to the center of a cluster of galaxies you find its central galaxy. Usually it's the largest and brightest galaxy in the cluster, but not always, because why should the universe be consistent. But it usually is.[2]

That central galaxy is usually a giant, amorphous blob of stars. You would be too, if you suffered countless collisions over the course of your life. A big, bright splotch. Not the most elegant creature in the universe, which is fitting, considering the monster in its heart.

While all galaxies host giant black holes, the central galaxies of a cluster will tend to be the home of the most massive monsters, having suffered collision after countless collision in their lifetimes, swelling both the bulk of the galaxy and the depths of the blackness in their centers. It's here where we'll tend to find the most fearsome activity, in the center of the center; one of the most dangerous regions in the known—and probably unknown—universe.

I've already talked about—in way too much detail, as usual—the centers of galaxies. The big hooplas. Well the central galaxies host the hooplas to end all hooplas. It doesn't get any bigger than this. Don't get me wrong: blazars and quasars can happen in any galaxy, anywhere, anytime, but the central galaxies are where the Real Fun is. Don't miss out.

As we dive into that central galaxy to examine the mystery and danger and mysterious danger of the quasar, passing the trillions of stars that live there, we come upon the first major inhabitant of the core: a slow-moving cold cloud filling up the thousand light-years or so around the central black hole. There are plenty of extra stars in this region too, packed cheek to jowl, a dense Manhattan compared to the sleepy suburbs of the galactic outskirts. Their combined light heats the gas, but gas is able to cool itself quickly, and its glow pales in comparison to what lies beneath.

But already, even at this distance, you can see—and feel—the radiance from the source. A distant brilliance, buried deep within a thousand light-years of gas, but unmistakable. Earth astronomers are able to see this from billions of light-years away; already we're far too close for comfort.

Pressing inward to a hundred light-years from the core, you come across the next layer: a cloud of molecules shaped like a donut, a tremendous pastry wrapped around the deep galactic core. The material here is surprisingly cool, emitting radiation and condensing on its own journey to the black hole. Its dusty clumps help to obscure and hide the light from the black hole, but you know you're getting close.

In fact, you're so deep in the core that you've crossed an invisible boundary: here the physics are dominated not by the forces of the host galaxy, but by the machinations of the central supermassive black hole itself. The gravity, the radiation, the intensity; the rest of the galaxy with its hundreds of billions of stars are completely forgotten. Here all ears turn toward the black hole and listen to its commands.

Ah yes, here we are. In the dragon's lair.

This is when the forces and energies become incomprehensibly large. Where the full fury of nature's wrath is evident and unashamed. Where life doesn't stand a chance. Where the insatiable pull of the black hole's gravity drives physics that defy description. I would say that this can only be seen to be understood, but frankly nobody has ever seen this environment up close—and lived to tell the rest of us.

The black hole itself may be full of mysteries, but in the end it's just simple gravity at work. But in this environment, in the home of the dragon itself, it's the full range of physics, from radiation and magnetism to heat and rotation. This is, without a doubt, the most powerful engine in the universe, driven and powered by the gravity of the giant void in the center.

Deeper you go, foolish and brave. Within a few light-days (no, I did not make up that term. You can have light-years, light-days, light-hours, even light-seconds) of the central black hole, you encounter the first signs of the enormous energies that these black holes are capable of. A hot, fast-moving cloud of gas, screaming as it winds its way to doom at the black hole.

The heat. Simple friction: the gravitational pull of the black hole is pulling a galaxy's worth of material into a region not much bigger than a solar system. It crushes together. As it falls, the simple molecules and atoms of the surrounding gas, once loose and free, cram together like too many people in a subway car. A confused, uncomfortable mass of bodies, sweating and waiting for it to end.

The light. Heated by friction, the gas glows, easily emitting more light than its host galaxy. Let this simple fact soak in for just a moment: as the surrounding gas falls into the central supermassive black hole, it heats up and emits enough light to *outshine an entire galaxy*. And not just once, but a million times over.[3]

The chaos. Flows of gas inroading from every direction of the host galaxy. Outflows pushing back. Radiation pressure. Flares sweeping up material. Turbulence and viscosity breaking apart clumps of gas and gluing them back together again. Buffeting even the most hardened ship like a toy.

It gets worse. Much, much worse.

Deeper you plunge, intent on seeing the very heart. Hidden behind the veiling cloud of thick, choking gas lies an even more violent region: the accretion disk. Exactly like what we see around newly formed stars and small black holes in binary systems, but scaled up to the nth degree. This is the final howling whirlwind of the gas before it finally plunges into the ever-hungry black hole. Heated to extreme temperatures, it spins at over a thousand kilometers per second.

Here, compression and friction rule the day, shaping the spherical blob of gas into a disk before it finally meets its end. It's easy to see why it forms a disk; it's the exact same physics that drives the collapse of solar systems. The spherical cloud of gas will be spinning, at least in some average way. Sure a blob or clump might have some random velocity in this direction or that, but on average the entire gas cloud will be rotating. It collapses, compressing from the intense gravity of the central black hole. The collapse happens in two direction: "in" and "down." That's kind of the definition of *collapse*. But the "in" direction is balanced by the centrifugal force of its own rotation, leaving only the "down" to survive. And so the infolding of a spinning cloud quickly forms a disk.

As it squeezes into an ever-tighter disk, the gas spins more rapidly. If you get caught in the disk, you are almost certainly doomed. While you're not quite through the never-go-back line of the event horizon, the enormous forces in the accretion disk compel you to follow so-called *tendex lines* (cool name for a band, by the way), following a twisting spiraling path further toward the monster at the center, drawn unflinchingly and without hesitation into the black hole.

All of this, the cool torus of molecules, the hot ball of gas near the core, the violent accretion disk, is all powered by one thing: gravity. The simple but uncompromising gravitational pull of the central supermassive black hole. The weight of a billion suns reaching out through the vastness of space to snare any pocket of gas into its deadly embrace.

You may never even make it through the event horizon, if indeed that was your goal. The temperatures and pressures and furious motions of the gas will do you in before you even catch a glimpse of that final edge. There is violence here, and it will visit terrible violence upon you.

The anatomy lesson isn't over quite yet, but to understand the next bit, we have to bring back an old reliable friend of ours: magnetic fields. "Like," you say in your stereotypical Valley Girl accent, "this dude is like obsessed with magnetic field, like, *totally*."

Wrong, bucko: it's the *universe* that's obsessed with magnetic fields. Totally.

I'm just the messenger. The universe is swarming with magnetic fields. All it takes is a little charge and a little motion. See that electron sitting on the bench, waiting for the bus? No magnetic field. It hops on the downtown express? Boom, instant magnetic field. And since the universe is full of charged particles and also full of motion, it's full of magnetic fields.

Usually these magnetic fields don't really do anything. Sure they may make your compass wiggle back and forth, but they're usually not strong enough to wiggle *you* back and forth. Usually. I've already talked about how they can snap like a crazy person and hurtle lumps of sun stuff into the solar system, and they're strong inside accretion disks, too. But it's not just a once-in-a-while buildup of magnetic tension. No, the fields here are strong and stay strong. A slow, persistent force that never tires, constantly renewed with fresh material and energy from the infalling matter.

The magnetic fields start out small but quickly—and I mean *quickly*—grow stronger. You ever hear of someone described as a "dynamo." Wow, that person never stops! Boundless energy, jumping from activity to activity without rest. Have you ever stopped to think about where that word comes from? It comes from physics, like almost all awesome words do. In the case of our intense little demon of an accretion disk, we're dealing with a *magnetic* dynamo. My favorite kind.

All that gas swirling and whirling to its doom is made of charged particles—it's too hot down there for neutral atoms, and all the electrons have been ripped off. Charged particles in motion equals magnetic fields, as I hope I've managed to explain already. The magnetic field they create follows the shape of the accretion disk, because . . . well, what else could it be? And the innermost regions of the accretion disk is orbiting faster than the outer parts. If you happen to be a world-class figure skater with a healthy interest in space exploration and astrophysics, then you already know why: pulling your arms in can turn a lazy spin into a speedy one. If you're not a world-class figure skater, then I'm sure you can use your imagination, or find a suitably greased office chair and give yourself a whirl.

The differences in rotation in the disk (called, not for the faint of heart, *differential rotation*) wind up the magnetic fields like wire on a spool. Like a string around a yo-yo. Like . . . I think you get the picture. It will take any old random magnetic field line and twist it up, strengthening its own total field.

But the flow of gas isn't exactly quiet and steady. No, you can image the chaos as gas collides with itself, meets resistance, fights back, raises plumes of hotter material, drops balls of slightly cooled clumps. It's messy. And if a big blob of gas happens to rise up out of the accretion disk, it carries with it a chunk of magnetic field lines. One part of this new lump of magnetic field will be closer to the black hole, and one will be further, because that's how randomness works. But the bit of field line that's anchored to the disk closer in has a faster orbit, doesn't it? It does, good job. That extra lump of field gets stretched, pulled, and eventually wound back down into the main field of the accretion disk.

And here's your dynamo at work: it converts random turbulent motion—energy in the accretion disk itself—into ever-stronger magnetic fields. As more gas finds its way onto the disk, the fields grow more powerful. Nifty.[4]

For extra deliciousness, here's the name of this particular effect: the *alpha-omega dynamo*. Sounds impressive, and people will think you're smart if you say it out loud.

Now here's why I care so much about these magnetic fields, and why you should care, too. When magnetic fields are weak, you move them. When they're strong, they move *you*. And boy can they carry a punch.

Remember those coronal mass ejections from the sun? And how powerful they were, able to knock out unprotected electronics in an electromagnetic storm? Those were absolute wimps compared to the powers and energies on display here. Blasts from the sun can knock out a satellite or temporarily overwhelm a planet's own protective field. But the magnetic energies driven by dynamos in accretion disks can do so much more.

As the rotating gas nears the black hole, the temperatures and densities increase rapidly. And so does the magnetic field strength. You can think of the field as a bundle of wires. This is, after all, how we like to draw magnetic fields: as lines. Do a random Internet search and I bet this is what you'll find. Closer to the black holes, those magnetic lines are bunched together, representing an incredibly strong field, wrapped around the midsection of the black hole.

And the black hole itself is spinning. It probably formed from a spinning star, and every time a new drop of gas falls in from the rotating accretion disk, it gets a little boost. Remember poor Alice, who got dragged into a spinning black hole? She got herself wound up like spaghetti on a fork, because of the innocent-sounding *frame dragging effect*: a spinning black hole literally pulls spacetime with it, forcing anything near the event horizon to rotate with it, even if it's not touching the event horizon. Nuts, I know, but that's black holes for you.[5]

What happened to Alice is now happening to these unfortunate magnetic field lines: dragged along with the spinning black hole. And then . . . something . . . happens. I wish I could say exactly what, but scientists are a little too primitive to fully understand the workings of this process. I can't blame them; it's a perfect storm of difficult physics. General relativity? Check. Strong magnetic fields? Check. Turbulent plasma? Check. It's enough to make even the brightest scientist crack.

There are a few ideas, named after the brave souls who attempted to figure these things out, such as the *Penrose mechanism* or the *Blandford-Znajek process*. I'm pretty sure these are curse words in some languages.[6]

But here's the gist: something in the combination of spinning black hole and strong magnetic fields in the encroaching accretion disk force the infalling gas *around* and *up*. First around the equator of the event horizon, following the tightly wound magnetic field lines. Then up along the very surface of the black hole, pushed up to the poles, and launched into tightly packed jet streams, thin columns of high-energy particles, free from the clutches of the monstrous black hole, screaming out in joy and terror as they rip across the light-years.

✪

These jets are signatures of a feeding black hole. Of an *active* galactic core. Twin beams, one launched from each pole, probing through the night. They are warnings of further dangers at their origins, and hazards in and of themselves.

It's hard to describe just how much energy these jets carry. All that pressure, all that tension, all that *weight* of the gas falling into the black hole. Some of that energy is lost forever, swallowed below the event horizon. But a fraction—and it only needs to be a tiny fraction; that's how tremendously powerful accretion disks are—escapes in the form of a jet. The particles that escape are mostly electrons, with a few positrons (aka the optimistic electron) and some protons thrown in for flavor. The jets reach dizzying heights: easily jumping past the accretion disk, punching through the surrounding clouds of hot and cold gas, and beyond. Past the core, escaping even the galaxy itself.

That feeding black hole is powerful enough to launch its twin jets light-years. *Thousands of light-years.* Compare that to the seemingly strong coronal mass ejections, which could only send a ball of gas limping halfway across a solar system. That's how hard, how fast, and how energetic these bad boys are. You don't mess around with a supermassive black hole. Certainly not near the event horizon, where its enormous and crushing gravity has some pretty nasty side effects.

At those energies, the particles in the jet are called *relativistic*. I know this chapter has lots and lots of technical jargon, but that shows you just how extreme these objects are: if an astronomical object is surrounded by a cloud of technical words, then you know it has to be special. Anyway, *relativistic* means that special relativity is important to properly describe their behavior. And that's a long-winded way of saying "really stinking fast." Like, a hair under the speed of light fast.

A spray of particles, traveling near the speed of light for thousands of light-years. They most definitely reach intergalactic space, but they can be seen even further. Those jets are glowing with radiation. Specifically, the radio kind of radiation. The jets are beams of fast-moving charged particles. No points will be awarded for correctly guessing what that means: a magnetic field! Hooray, those again. Just what we needed.[7]

In fact those jets are beamed and shaped by a protective sheath of magnetic fields, keeping them straight and narrow across the vastness of empty space. Those lines wrap and coil around the beam, forcing the particles to travel straight paths down the barrel on their journey. And there's a bonus feature to fast, charged, slightly curving particles: emission of radiation. Noticed by Earth scientists when they first turned on *synchrotrons*, which are machines for speeding up and slightly curving charged particles. Hence, *synchrotron radiation*.[8]

I know, I know. More jargon words. Look, I'm enriching your world and vocabulary. You can pretend to be a smarter person. You'll thank me later.

Assuming you survive.

This radiation tends to come out in radio wavelengths, and can be seen from the other side of the universe—it's this radio emission that the early Earth astronomers first identified as dots scattered around the sky. Quite literally: we see jets and light from their hosts from billions of light-years. I've spoken about supernovae, which also outshine their host galaxies. But whereas those explosions fade after a few days or weeks, these active galaxies persist for millions of years.

But the radio isn't all there is. Since the particles in the jet are moving so quickly, if they happen upon a low-energy photon, say some infrared light, they can give it a good swift kick, turning it into a lethal high-energy X-ray or gamma ray photon. And since there's a lot of particles and a whole lot of jet, this happens a lot. So there you have it: high-energy particles, high-intensity radiation, millions of times stronger and brighter than anything else you're likely to see in this universe.

A fiercely burning lamp, refusing to extinguish for millennia, casting its rays, calling from one galaxy to another.

✪

These quasars emit more light, more energy than every star in their host galaxies combined, a million times over. And they can do it unceasingly for millions of years. You don't even have to be near it to feel its effects: get one of those jets blasting away in your face and soon enough you won't have a face to be blasted with. And you certainly don't want to cruise near the core: get caught up in the violence of the accretion disk, and you'll be swirling away with the rest of it, twisted and reshaped before meeting a grisly fate in the black hole or an even grislier one ejected from it.

How much does the black hole need to eat to power such an inferno? A typical moderately powerful midsize family-friendly quasar will consume a few hundred suns' worth of stuff every year, which works out to a few hundred Earths *every single minute*. And you thought you had an appetite. Only a few percent of the energy generated from the falling gas is needed to generate that intense heat and light.

In the end it's just gravity doing the work: water falls down a hill, and you can attach a wheel with buckets to it and gets some of the energy out. Gas falls down to the center of a galaxy and into a black hole, and some of that energy gets transformed into fuel for the galaxy-hopping twin jets.

I should take a break from scaring you and talk about the names. From the moment Earth astronomers discovered them, they knew something was fishy. They were looking at objects brighter than a galaxy, but with no discernable shape. Lacking any other convenient metaphors, they decided they looked kinda sorta star-like, hence "quasistellar object." OK, fair enough, we'll give them that.

But that's too long-winded, so they shortened it to *quasar*. Hmmm, a little weird, but none of my business. And then a whole host of somewhat-related objects were discovered and given colorful names like blazars, Seyferts, and LINERS. Don't ask me what LINER stands for, you don't want to know.

After some significant head-scratching astronomers figured out that they were all really the same kind of thing: a galaxy hosting an active core, or active galactic nucleus. What were once thought to be different objects, and hence all had different names and labels, were realized simply to be different views of the same kind of creature. Some active galaxies are simply more active than others. Some are louder or quieter on the radio. Some put out copious amounts of X-rays. Sometimes we're viewing the action through the dirty torus surrounding the black hole. Sometimes we're staring right down the barrel of the jet itself.

No matter if it's a quasar, a blazar, a Seyfert, or anything else, it's always due to the infernal machinations of the nucleus of an active galaxy.

Now, back to scaring.

Here's an example of just how powerful quasars are. Quasars live inside galaxies, which cluster together in clusters of galaxies. But most of the stuff in a cluster isn't beaded up in little galaxies, but just floating around unbound. Oh yeah, there's also dark matter, but that gets its own chapter.

The galaxies are the beans and the gas is the broth in this giant soup. And just like soup, it's slowly cooling off. As it cools, is condenses, pooling into the inner core. But there the central galaxy and its mighty black hole await, ready for more food. The additional gas feeds into the accretion disk, lighting it up and launching the jets. The jets, much hotter and much livelier than the cooling gas in the outskirts of the galaxy, heat it up.

And this is the cool part: the jets are so well controlled for so long that they act like gigantic straws, and blow bubbles of hot plasma all the way into gas that makes up the cluster of galaxies. Just like you sticking a straw in a drink and blowing some bubbles, annoying your parents and making a mess. Except these bubbles are thousands of light-years long and take millions of years to inflate.

These bubbles rise up into the cluster atmosphere like a hot-air balloon, where they eventually pop, dispersing their contents and letting them mix into the general cluster atmosphere (named, for the nerdily curious, the *intracluster medium*). The surrounding gas, now much warmer due to the injection of all that gas, is a little more cautious about getting near the black hole. Starved, the great engines shut down: the accretion disk slows, the jets turn off. Eventually, after millions of years, the gas cools once again, stoking the fires yet again.

This constant heating potentially solves a little puzzle in astrophysics. That gas in the center of a cluster is good at cooling itself off—too good, in fact. Simple back-of-the-envelope calculations estimate that the central gas should have cooled itself off billions of years ago and clumped together to form a massive, hideous galaxy. And while the galaxies in the centers of the clusters are indeed big, they're not *that* big. At least, not as big as they should be, given how efficiently the gas around it can cool off.

But with the energies released from the quasar, the gas can stay warm. And anytime it feels like chilling out and condensing onto the galaxy, more gas flows inward to the waiting black hole, which compresses onto a now-enlarged accretion disk, which glows with insane energies and radiation, some of that gas swirling around and up into a jet, heating up the gas that is trying to find its way down to the core.

We can see the effects of this: trains of bubbles, one after the other, blown out by the central black hole. A heartbeat for every galaxy cluster, with a rhythm millions of years long.

This feedback can potentially keep the gas in a cluster of galaxies warm for billions of years, through the constant repeated on-off cycling of the quasars and their jets. Imagine that: an engine powerful enough to warm an entire cluster of galaxies, to keep the fires lit warm enough to heat a hundred thousand galaxies' worth of gas, spread out over a volume a million light-years across.[9]

And the effects of these awesome energies aren't just felt in the larger clusters, but in the galaxies themselves. Just image the ferocious power washing over you, helpless and in awe, as one of these jets activates or the quasar lights up. Stars form from the collapse of gas, and gas needs to cool in order to collapse. It seems that the lives of all galaxies—not just the central ones in a cluster—are governed by the activity powered by their giant black holes. Gas cools and condenses, forming new batches of stars throughout the galaxy? Good, great! But some of the gas will condense onto the black hole, driving a quasar event, pushing some gas out of the galaxy altogether like the bully that it is, and heating up whatever remains, choking off star formation, until that gas has enough time to cool and collect itself once again.

This feedback makes itself known through a peculiar relationship between the mass of the central black holes and the measured properties of their host galaxies—it seems that galaxies and their black holes don't just live together, but *coevolve*. It's not a one-way relationship. The bigger the galaxy, the bigger its black hole, true, but also the bigger the black hole the bigger its host galaxy. They are tied together, inexorably, through the course of billions of years of cosmic history. The galaxy feeds the black hole, the black hole heats the galaxy, preventing that black hole from consuming all the available gas in a single fit of rage-induced hunger.

Back and forth, forth and back, the galaxy and its giant black heart trade energies and forces, self-regulating and self-moderating.

The central black hole is just a tiny percentage of the volume and mass of its host galaxy, but the unparalleled gravitational energies demanded by the very existence of that flaw in spacetime are able to govern, shape, and sculpt their hosts, for millions and billions of years.

Indeed, galaxies may have burned up all their material in the form of stars long, long ago if it weren't for the feedback energies of their central black holes. A galaxy, unregulated, may simply use up its supply of available material in the first billion years of its life, quickly succumbing to the dull retirement of red, dead, decaying stars for the rest of the future history of the universe. But the energies released by the black holes slow down its consumption and conversion of material into stars, enabling the galaxies to shine for trillions of years; enabling generation after generation of stars to live and die, each new round slightly richer in heavy metals than the parents that came before.

In turn, the galaxies are able to regulate their supermassive black holes, only allowing it to sip rather than gorge, keeping it from engulfing the entire galaxy.

The sun, the Earth, life, and you yourself might owe your very existence to these fearsome energies, and the intimate connection between bright galaxies and their black holes.[10]

✪

As awesome and fearsome as quasars can be, thankfully, the universe is a relatively quiet place nowadays. Most of the quasars we see are distant and old, a relic of the cramped and chaotic early universe. In fact, the discovery of quasars was one of the first indications that the big bang model of the universe was correct. While the fine details are beyond the needs of your average traveler, the general picture painted by the big bang is that the universe changes with time (a wildly radical suggestion when it was first proposed at the beginning of the 20th century).

When astronomers first spotted quasars and realized that they exist outside the Milky Way, they calculated their distance and found them to be exclusively *really far away*. The nearby universe doesn't host a single active galaxy.

Astronomy is a time machine. It takes time for light to hop from star to star and galaxy to galaxy. By the time the light reaches our eyeballs and our telescopes, we don't get an image of a distant object as it is *right now*, but as it was when it sent the light to us. It could be eight minutes ago; it could be eight billion years ago. So the further out in space we look, the further back in time we perceive.

The fact that quasars appear in only the distant universe means that they are a feature of only the young universe, which is a pretty clear indication that Times Back Then were different—the universe has changed character and flavor over the epochs, a key prediction of the big bang. It's virtually impossible (at least, without a lot of twisted mathematics and tortured logic) to come up with a model of the universe that keeps it eternal, yet changing.[11]

Quasars were awesome forces to contend with . . . billions of years into the past. Today? Quiet—perhaps too quiet.

Most galaxies today are relatively calm. Sure, they still host destructive potential with their black hole hearts, but with nothing to feed them—especially not hundreds of Earths every second—there's no accretion, no jets, no danger.

Still, there's always *some* amount of material nearby the supermassive black holes. This is how we discovered Sagittarius A* in our own Milky Way: through the copious radio emission from the disk of material around it. Black holes almost always feed, just not on the level necessary to label their host galaxies as active. But don't be fooled by their relative quiescence; they're still dangerous and generally inhospitable and just downright unfriendly.

Even Sagittarius A* has a jet—a small one, but one nonetheless—and as far as we can tell, it's actually pointed in the direction of the Earth, our home world. It's not enough to pose a danger to life on that planet . . . assuming that our black heart doesn't find any opportunity to consume a buffet of gas in one sitting and blare out a furious beam of radiation into this portion of the galaxy.

But it doesn't take a giant feast to make a black hole hostile. Remember when we ventured far too close to Sagittarius A*, and found a group of daredevils orbiting too close to the event horizon? Sometimes those foolish stars get close . . . too close.

You can put just about anything in orbit around just about anything else without anything interesting happening, but like all things there are limits. The gravitational forces that give rise to the tides can be nothing more than a minor disturbance—as the Earth finds with its gentle rising and falling of ocean waters—or it can be a disaster—as Alice found as she became spaghettified on her approach to a black hole.

Sometimes those daredevil stars get too close, and the tidal forces are enough to overwhelm their own gravitational integrity, ripping those stars apart at the seams. That cloud of gas—which was once a noble star—quickly compresses into an accretion disk and drives itself crazy with tortuous magnetic fields. Most of the gas falls to its doom beneath the gaping maw of the event horizon, but some is able to find freedom and safety in the form of a brief flash of radiation and particles in a jet.

✪

It's suspected that almost every galaxy went through a young phase of *activity*, but most today are middle-aged, and have moved on from such violent youthful indiscretions.

Mostly. We're not exactly sure what activates a quasar. It takes a lot of material, moving quickly inward, to trigger such a massive event. One suspected cause is that flare-ups tend to be triggered by the mergers of galaxies, like a bad case of indigestion after a heavy meal. With a galactic merger comes a fresh supply of gas hurtling toward the core and the now-larger black hole. Feeding time.

This would explain why quasars are more common in the distant universe. When the cosmos was younger, it was smaller, with the same amount of material crammed into less volume. With less room to maneuver, galaxy collisions were far more common than they are today. And with those mergers came violence, activity, larger black holes, and enough gas to trigger the formation of a quasar.

Nowadays, with things more spread out, the galaxies just aren't as active anymore. Even star formation itself is a mere shadow of what it was billions of years ago. It's sad when you think about it—our universe is already dying—but we should be grateful. With less activity comes fewer quasars, which means life has a better chance to staying alive than it did in the young universe.

But new quasars can still potentially form, especially after a merger event. And don't you think that our home galaxy is immune to the carnage. We are, after all, on a collision course with our nearest neighbor, the Andromeda. When we finally begin to mix in a few billion years, a large portion of gas will find its way swirling to the core. Our black holes will merge together. And it will be hungry.

When it feeds, the union of our galaxies may find itself host to a powerful new quasar, irradiating and eradicating any unprotected life nearby.

And it looks like this has happened before. Astronomers searching with gamma-ray telescopes found two giant bubbles of thin but incredibly hot plasma, extending for 25,000 light-years on each side beyond the plane of the galactic disk. Called the *Fermi bubbles*, we suspect that they are the faint remnant of a feeding frenzy by Sagittarius A* from millions of years ago. What was our home galaxy like when that cataclysm took place? Did any life that gained a foothold on some distant world back then suffer from being too close to the blast?

Do we really want to find out?

Thankfully quasars are mostly a relic of the youthful universe. An artifact of a bygone age. But despite their distance their voices can still be heard, billions of light-years away and billions of years later. They shouted once but loudly, refusing to be ignored for generations to come.

A galaxy like our own that appears quiet—perhaps a nice place to settle down, find a patch of land in the suburbs to call home—can host deadly surprises. All it takes is a fresh batch of material to find its unlucky way to galactic core. Once ensnared by the black hole, it follows the same terrible trajectory as its ancestors did long ago. And with every new appetizer comes a new ejection. A new flare, a new jet, a new blast of lethal radiation and particles.

Active galaxies are highly variable, too. You might come upon one that you think has just shut off, wiping the last bits of food from its fat lips. But within even a few days it can light back up again, with its jet pointed right at your unsuspecting ship. These short but fierce blasts may be the explanation for the generation of the highest-energy cosmic rays ever discovered. Rare, thankfully, but not unheard of. It's a chaotic and random universe out there, and what appears to be a safe harbor can be anything but.

Here's a good general rule for intergalactic space travel: if can see a fire blazing from the next galaxy over, *don't go and check it out*. See, that's not so hard, is it?

PART FOUR

SPECULATIVE THREATS

Cosmic Strings and Miscellaneous Spacetime Defects

You broke my heart
This one string
plays but a single note:
It's a big universe.
　　　　—Rhyme of the Ancient Astronomer

The birth and death of stars. Storms of cosmic rays. Black holes great and small. All very real threats, but relatively new on the cosmic scene—even those that are billions of years old. Everything we've encountered so far is a byproduct of the natural life and evolution of matter in our universe. Collapsing, fusing, processing, changing through the eons from one form to another, and generally making our lives miserable in the process.

But some dangers are far older. Relics of the ancient universe. Unchanging since the big bang itself. Created before the first generation of stars, before even the first atoms coalesced. These old dangers are not sprung from fusion or magnetic fields. They are not birthed in the clouds surrounding black holes or the remnants of dead suns.

No, they were forged in the fundamental physics that ruled the earliest moments of the universe, in the intense heat and pressure where disparate forces were united. Where bizarre fields flared briefly to reshape the destiny of the universe, never to be seen again.

The earliest moments of the universe were ruled by physical laws never to be encountered again. Mere microseconds after the big bang—the singularity before which we cannot probe with theory or experiment—the entire universe could fit in your hand but was too hot to hold. Only there were the conditions right to create such monsters.

They are defects in the very fabric of spacetime. Flaws in the otherwise beautiful and smooth landscape on which we live. Regions of crushing gravity and exotic forces. They roam freely throughout the vastness of the cosmos, ruining stars and cleaving galaxies. Like the demons of ancient lore, they are powerful, but rare. Perhaps only one or two survive to the present day. But there they lurk in the shadows.

If you want to slay dragons, you may hunt them, but remember: every dragon's cave is littered with the bones of failed knights.

★

Before we get started, we need to lay down some facts of the universe. I know, you didn't buy this book expecting to read a lecture, and I certainly didn't *write* this book expecting to deliver a lecture. But sometimes you have to drive through a day's worth of cornfields to get to the city, and sometimes you've got to learn a few facts before getting to the juicy bits.

Let's not waste any time:

Fact #1: The universe used to be a lot smaller and a lot hotter. It's not hard to figure out, once you discover that every galaxy is moving away from every other galaxy. On average, of course. Sometimes a galaxy will crash into another galaxy because of their local gravitational attraction. The Andromeda galaxy is headed on a collision course for the Milky Way, for example, but like I said earlier, we've got 4 billion years plus change before we get to that cosmic car crash, so don't sweat it.

But overall, the whole place is getting bigger. Which means in the past it had to be smaller. Big leap of logic there. Who said physics was hard?

If you take the same amount of stuff and cram in into a smaller space, it's denser. That's kind of the definition of density. And stuff with a higher density tends to be hotter.

Put these ideas together, and if we see a big, cold universe today, then a long time ago it was small and hot. How small and hot? How about this: at one important state, less than a second into the big bang, the entire universe was the size of a peach and had a temperature of over a quadrillion degrees. See, I told you so: small and hot.

You might start wondering just how small and just how hot the universe managed to get. But at some point we reach the same problem as we do at the center of a black hole: technically there's a singularity at the beginning of the universe, a point of infinite density and zero size, but we know we're doing the physics wrong at those scales, so let's not worry about that right now. Besides, we don't even know if the word "beginning" makes sense, seeing as how we're discussing a complete and total breakdown of all known physics, including the physics describing the relationship between space and time.

But that's not our problem.

We know this picture is pretty much correct, not just because it's based on naked facts, like the observed expansion of the universe, but because this idea of a small hot young universe gives several *predictions*. You know, how science works and all. The afterglow light pattern from the early cosmos. The balance of hydrogen and helium. So on and so forth. And wouldn't you know it, this picture pans out.[1]

Fact #2: When things—water, meat, universes—change temperatures and pressures, they can change *phases*. "Phase" is just a formal way of saying "state," which is a business casual way of saying "a way that stuff can be organized." Solid. Liquid. Gas. Plasma. The usual suspects.

Every bit of matter in our universe is in some sort of state, and given the right conditions those bits of matter can change from one state to another. Usually this change is rapid and uncomfortable for the bits of matter involved. Heat up liquid water and it changes state into a gas. Freeze carbon dioxide and it becomes a solid. Phase changes. You get the idea.

You may be wondering why I'm pointing out something so obvious. That's because . . .

Fact #3: Subatomic particles aren't what you think they are. For instance, you might think they're particles. "Here's a bag of electrons, I got them for your birthday," you might say, as you shake a teensy little bag of subatomic marbles. Even I'm guilty of using the metaphor, describing high-energy cosmic rays as tiny little bullets, for example.

And that metaphor is useful and appropriate, in some cases. After all, why bother calling them "subatomic particles" if they don't act like particles? That's because they're not particles . . . always. Sometimes they act like waves. And sometimes both. It's . . . it's complicated, all right? Some aspects of the physical universe are just going to be a bit messy.

But here's what matters: don't think of individual particles as individual particles. You think you've got an electron, I've got an electron, and they're unrelated electrons. They just happen to have the *exact same* charge, mass, spin, and all the other properties. Have you ever sat down to think about it? Why is every electron exactly the same as every other electron? In fact, they're so much the same that we can just call them all "electrons" without having to name individual ones. We don't need Harold Electron and Wanda Electron.

Hm. What if all electrons were a part of something bigger? What if there was something that spread throughout spacetime and connected all the electrons together, man? OK, not a bad idea, let's run with it. Let's replace "all these electrons running around in the universe" with a single "electron field." An electron essence that permeates the universe. That thing, the field, carries all the properties that make an electron an electron. And to get an individual particle that we can toss around, we just . . . pinch off a piece of that universal field for our own personal amusement.

It sounds trippy, and it is. But this is the very heart of a fundamental theory of how things work: *quantum field theory*.[2] All the particles and forces that we know and love—electrons, quarks, photons—are really discrete manifestations of larger continuous fields that exists throughout spacetime—the same fields that we met when we were poking and prodding at the nature of the vacuum. One field for each kind of particle. The concept of "pinching off" particles from the field isn't 100 percent accurate, but it's a good enough analogy to get you the flavor of what's going on.

OK, fair enough, so what? It means that the universe is full of these fields. Remember when we looked at the exotic nature of the vacuum of spacetime all the way back when we first left the comforting embrace of the Earth? It's the exact same thing: a vacuum really isn't empty, it's filled with a bunch of fields, just no particles have been pinched off in that particular patch of the field.

Back to our fields. As the universe gets smaller or bigger, the fields contained in it can change their phase, just like anything else. One more piece in the puzzle.

Fact #4: The fundamental forces of nature aren't what you think they are, either. Gravity. Electromagnetism. Strong and brave nuclear. Weak and puny nuclear. I push you, that's electromagnetism at work. You fall to the ground, crying, that's gravity.

In the quantum field theory picture of the universe, though, they're just so much more field. An electromagnetic field. A gravitational field. And you can pinch particles off them, too. Take a piece of an electromagnetic field, and what do you call it? A photon, that's what. As in, light. Huh, isn't that interesting.

So particles are fields, and forces are fields too, all saturating the universe like so much olive oil and balsamic vinegar in a chunk of bread.

Here's the wild bit: all four forces, which couldn't be more different from each other, might all be facets of the *same* force. Electromagnetism, gravity, strong nuclear, and weak nuclear might be all aspects of a single force, a single field. It sounds nuts, and if you start spouting it on the street corner, you will rightly get a bunch of funny looks. But this idea kinda sorta works: increase the temperatures and energies enough, and forces start to . . . combine. A cosmic Voltron, if you will.[3]

The easiest to pair up are electromagnetism and the weak nuclear force. At high energies, they're not different entities anymore. There's only a single force: the *electroweak force*. Cool name, isn't it? It's only at low energies (and by "low" I mean normal everyday energies) that they appear to be two separate forces.

Particle physicists go nuts for the word *symmetry*. Whisper it in their ear and they'll love you forever. They say that at high energies the unified force is *symmetric*, but at low energies the symmetry "breaks," leading to two different forces. There are miles of math to back that up, if you're masochistic enough.

It seems weird to us that these two forces, with very different ranges, strength, and sources of interaction, could really be two sides of the same coin. The electromagnetic force has a single carrier—the handsome photon—with infinite range, and does all sorts of interesting things from illuminating the cosmos to holding your kids' pictures up on the fridge. The weak force has three dumpy carriers (the W^+, W^-, and Z bosons, in case you're curious about their names) and has incredibly short range. The weak force does have a cool superpower—it can talk to neutrinos and can change one kind of quark into another—but that's basically it.

How could these two things possibly be related?

But we live in a broken universe. Look at a cracked mirror: you see two different reflections. Could you ever believe that those reflections were the same, once upon a time? At high energies the photon disappears, along with the three carriers of the weak force. In their place are a quartet of massless particles responsible for carrying the electroweak force. And when the temperatures get too low, this beautiful symmetry snaps, forcing three of the carriers to become massive (which gives rise to all the physics of the weak force) and liberating the photon to roam free throughout the universe.[4]

At even higher energies the strong nuclear force joins the unification party (because there's yet another symmetry at play there), and beyond that even gravity gets in on the action. Supposedly—we actually haven't figured out a way to get gravity to play nice at high energies, hence all the question marks surrounding singularities and black holes. The search for a quantum theory of gravity is a hunt for a super-duper symmetry that explains it all.

Wow. I know I've typed way too many words without a ✪ to break them up. You must be tired. Get a glass of water, take a nap. Go for a walk. I know I'm about to.

I'm doing this for a reason, and that reason isn't, "I need to write a minimum number of words for this chapter or my publisher won't let me include it." The dangers I'm about to describe are . . . exotic. They're not made of the normal stuff of space. They're not waves of particles or radiation. They're not highly compressed normal matter. They're certainly not lumps of rock and ice.

The hazards I'm about to describe are leftovers from an era when the universe was fundamentally different than it is today. From when the universe was *organized* differently. The physics of the first instants of the big bang were literally different than they are today. And what those physics produced—and may have left behind—are strange beasts, and to describe them properly and to give a full account of their terror, I have to dig into the physics.

The physics of these forthcoming menaces rests on the combination of the four facts I just described. Taken together, the story goes like this: A long time ago, the universe was smaller and hotter. The fields that make up the fundamental forces and particles were in a more unified state, where the universe played by a different set of rules. As the universe cooled, the fields experienced a phase transition and the forces broke apart. Havoc ensued. I couldn't just say all that without backing it up, but you know me better than that. I don't pull punches when it's your life on the line.

Good, now that we've got the pregame warm-up out of the way, let's play ball.

★

In the very earliest moments of the universe, things were much more unified than they were now. There wasn't electricity over here, pushing against charges and pulling on currents, and strong nuclear over there, gluing atomic nuclei together. There was just the Force.

We think. I mean, the math is a bit difficult, even for people who spend their day jobs worrying about this kind of thing. And when scientists get worried, they attach a name with a cute acronym to it. In this case, we have GUTs: grand unified theories, or theories that try to combine electromagnetism with the two nuclear forces at very high energies.

What about gravity? If you toss that in, you don't have a GUT, you have a TOE: a theory of everything. I know, I know, horrible stuff, and if I had a time machine the second or third thing I would do is go back and slap those Earth scientists for coming up with such eye-rolling names.

Anyway, time-travelling-scientist-slapping fantasies aside, let's look at the GUT era in our universe. What was it like?

Hot, mostly. And cramped. And weird. Like that one party you went to in college.

In this hot cramped young universe, life was different. Under the extreme pressures and temperatures, and subject to the strange unified forces, strange particles were created and destroyed in the flash of an eye. These particles could only exist in the peculiar conditions of the early universe, a byproduct of the combination of forces. Today, there just aren't enough places in our cold, empty universe that have the right conditions for creating these particles anew. The most powerful particle colliders in the world don't come within spitting distance of the energies needed—even a collider looping around Jupiter wouldn't have a billionth of a billionth of the energies needed.

The only way we can hope to understand this epoch of the universe—which lasted less than a tiny fraction of a second, by the way—is through the hopes and dreams of our chalkboards and laptops. And through those dimly lit searches we struggle to find a way to test those ideas against some form of low-energy experiment (maybe a new particle is predicted to exist, or some tiny modification to a decay rate, and so on).

Still, physicists have developed several contender GUT theories, and since they're the professionals that's all we've got to work with.[5]

The force of the universe in its initial moments was a frightening, chaotic mess: these strange species of particles were created, absorbed, reflected, destroyed, intersected, trapped, and excited from moment to moment. Almost all were certainly destroyed, as the processes that created them were in equal ferocity to the ones that obliterated them.

Did I mention that at this time, the GUT era of the universe, our entire observable cosmos was smaller than an atom? Yeah, intense.[6]

Surely almost all the exotic particles of the early universe were unstable, with fleeting lifetimes even shorter than the length of the GUT era itself. They would have simply fizzled out into showers of more familiar particles as soon as they got the chance.

But some of the strange particles crafted in the GUT era may have survived. They may have avoided capture or eradication. They may have survived the subsequent expansion and cooling of the universe. Hidden in the shadows. Streaming from one end of the universe to the other, waiting for a chance to strike.

Bizarre and ancient particles, impossible to re-create in the laboratory today. Leftovers from an age that the modern universe has forgotten. Some of them even have names.

One is known as the *monopole*.

Doesn't sound so scary, does it? It could've been *monsterpole* or *disaster-pole*, but *monopole* just seems so . . . academic.[7]

Here's why they're weird. You know what magnets are, right? You've got some in your house. The Earth has a big one. One end is called "north" and the other called "south," for convenience. What happens if you sawed that magnet in half? Would you have a north piece and a south piece?

Since I'm asking the question, you probably know that the answer is whatever's weirdest. In this case, you don't get two half-magnets, you get two smaller north-south magnets.

Do it again, and again, and again. All you'll end up with are a bunch of minimagnets, each with its own tiny little north and tiny little south.

When the universe behaves in a counterintuitive way, that's its way of whispering its secrets to you. If you listen carefully and think hard enough, you can divine those mysteries.

Ahem, sorry to get all mystical on you there, but you get the point.

What the universe is telling in this case is that magnetism isn't really an *intrinsic* property of stuff. It's different from, say, mass or charge or location. It always comes about from the *movement* of charges. Another way to say that is that electric charge is something that can happen by itself, but magnetism isn't.

Inside your lump of a magnet you have lots of electrons whirring around. Each one is spinning, orbiting, and doing whatever electrons do when we're not looking. But, essentially, they're moving. And that movement creates a magnetic field, with a north and south pole. Chop up the magnet, and you still have a block of electrons, creating the two poles.

Always north and south poles. Two poles. Let's pretend to be scientific and use a Greek prefix and call it a *dipole*.

As far as we know, going back for hundreds of years in experiments, magnets can't come any simpler than that. There's no such thing as "magnetic charge" the same way as there's "electric charge." You can have positive and negative charges together, or you can have them separate. You can, for example, have a single solitary electron, sitting there alone, and it will have an electric charge of negative one (by definition, thanks to none other than Benjamin Franklin).

A charge by itself. One charge. A *monopole*.

There are electric monopoles, but no magnetic monopoles. No solitary north poles floating around. No unique south poles that you can play with.

That is, there are no monopoles *today*.[8]

The early universe was a different story. And that story, suffering as it was under the weight of a unified force, could produce unusual particles. Particles with, say, a magnetic charge.

Particles carrying a single north or south pole with them, just like they could carry an electric charge or a mass.

Good luck making one in your at-home E-Z chemistry kit, but GUT models actually predict loads of them in the early universe. In those days, spacetime itself was a seething, frothing nightmare, due to the incredible energies of the quantum fields. As the universe aged and cooled, most of that boiling reduced to a nice calm simmer, but in some places spacetime itself got, for lack of a better word, stuck.

Particles can stay stable as long as there's nothing lighter that they can decay into (NB this is subject to various rules of quantum mechanics that we won't dig through). The particles of our everyday existence are the lightest of the ones in the zoo; there's nothing else for them to transform to, so they stay where they are. The monopoles produced in the GUT era are knots in spacetime that remain exactly as they are.

Back in the day, the universe had enough energy to pop out these kinks with a good massage. But nowadays, the cosmos is far too old and cold—there simply isn't enough energy going around to undo the leftovers of the GUT era, and so the monopoles persist.

It's a shame most of them got inflated away.

Inflated? Oh, yeah. It's not a threat (anymore) but it is thought to play a major role in the early universe. Rough sketch: there was another field hanging out at the time (called, get this, the *inflaton field*. Gag me.), it slipped, and caused the universe to rapidly expand in the barest, tiniest hint of a second. And I don't mean just a little bit expand—the universe, in that flash, became at least 10^{62} times bigger than it was.

Again, intense.[9]

Once cooled off, that inflaton field decayed into all the particles we know and love today. Lots more to this story, of course, but that's another book. A book on good things in the universe and our place in it, maybe.

So the upshot is that there used to be monopoles everywhere; you couldn't walk the dog without it chasing after one, and all those monopoles were constantly popping into and out of existence. Some of those monopoles got locked in place as the universe aged, but then the whole thing went *whoosh*, and in our observable patch there's only a few remaining, at best.

Nobody's exactly sure what it would be like to encounter a monopole. It depends on which GUT theory—if any of them—turns out to be right. They're most likely unstable. Unstable in the radioactive sense, I should clarify, and only over very long timescales. They can interact with light, so you might be able to see them coming. Or not. They might harmlessly pass through you, or they might carry so much momentum they could wreck you up.

Hard to tell. Check back in a hundred years or so once those scientists have come to grips with the whole issue.

That said, they're certainly incredibly massive and incredibly exotic. And do you want to come face-to-face with something massive and exotic and massively exotic? I didn't think so. And they can either hurt you directly or by—surprise!—decaying into a shower of high-energy particles, just like they were able to do back in the good old days, before inflation.

Maybe there's a clutch of them hiding somewhere in the cosmos. Maybe by now—13 billion years since that primordial turmoil—there's only a single monopole left. And you might just be the lucky boy or girl who gets to encounter one. That's part of the magic of space travel: you never know when an unprepared explorer might just become a one-of-a-kind science experiment.

✪

Monopoles aren't the only members of the early universe freak show cast. In fact, you could argue that they're the most normal of all of them.

Remember that bit a few pages back about phase transitions? And how the universe went through a few of them? Let's stick our thumbs into this phase transition pie and see what we find.

Fill up a glass with water (liquid, please) and stick it in the freezer. Or if you happen to live on a suitably cold planet, outside. Twiddle your thumbs for a while. Check on the water. If you wait long enough, depending on the amount of water and just how cold your freezer is, eventually you'll have a glass of ice. It's still water—a couple Hs chilling with an O—but in a different state.

As a liquid, the water molecules were all wishy-washy, flopping here and there wherever their little carefree hearts would take them. A liquid. But the freezer sent them to a hard knocks reform school, and as an ice they're all marching in rigid lockstep, forming crystal patterns all lined up the same way. A solid.

What happened? A phase transition happened.

Heat the water up again, and the molecules get enough energy to start having rebellious thoughts and start breaking ranks. The crystal structure breaks down, and the phase relapses right back into a liquid, its former regimented self completely forgotten.

But take a look at that glass of ice again. Look deep inside. I'm serious: This is an experiment that's safe enough to do at home. Chances are that piece of ice won't be perfectly, um, *crystal* clear. Inside the ice there will be cracks and walls and other deformities.

Ice doesn't *appear* from water. It's not magic, it's physics. To make ice, the molecules have to line up, and they have to start lining up somewhere in the water. Once that lineup gets going, it serves as a seed point where nearby water molecules can latch onto and join ranks. Fair enough, no big deal. The ice starts to form at a certain point and spreads throughout the water.

But what if two points got started at the same time? The transformation from liquid to ice doesn't happen instantaneously—it takes time for the liquid-to-ice transition to spread from a seed point through the rest of the volume. And in that time more than one seed could ignite the transition.

Eventually the two ice-forming regions will meet, like expanding old world empires.

And where they meet, there's conflict. When an ice seed forms, the molecules just pick one random direction to line up their crystals. Say, for example, up and down. Everybody that joins that party is forced into the up-down lineup. But another seed in another part of the water will have completely different random direction. Say, left and right. And any water joining *that* seed will be forced to be left-right.

All the water ends up joining one or the other party. And where they meet there's a transition line from up-down land to left-right land. A barrier. A flaw. A defect.

So it goes for water, and so it goes for the universe.

Except, you know, it's a lot hotter and a lot weirder. And it doesn't involve water. Instead the things doing the phase transitioning are the fundamental fields and forces that make up the physics of the universe. As the universe expanded and cooled in its earliest epoch, the fields froze out. Today, in the present epoch, we identify those frozen-in fields as the split forces of nature.

Liquid water has a certain symmetry—it doesn't matter which direction you look at water, it's just water. But a lump of ice looks very different depending on the way you look at it. The original symmetry in the liquid has been lost, and a particular choice (or choices, depending on how many seek points you have) of orientation have been selected out. The symmetry has been broken.

When our universe cooled within its first second, the elegant and beautiful symmetry of the unified forces froze out, and a particular choice was made; that choice being the forces of nature remaining today.

And just like with water and ice, some portions of the field froze out in a particular configuration, and some portions in another. It's still ice, no matter what, but a slightly different configuration of that ice. C'est la vie, I suppose, because nothing's perfect, especially in this universe.

And where do those domains in ice meet? A flaw, a crack, a wall.

✪

When the universe cooled, different parts of the universe may have cooled in a slightly different way. The entire universe still ends up with the same forces and fields no matter what, but their fundamental structure could be oriented differently from place to place.

And where do those domains in spacetime meet? A flaw, a crack, a wall.

A cosmological defect.[10]

A bend, a wrinkle in the very fabric of spacetime. The most common form is a one-dimensional line: a *cosmic string*. For those of you with too much time on your hands, this is not to be confused with *superstrings*, which are a candidate for a TOE known as string theory (although it is possible in some theories that superstrings can become cosmic string). But let's just move on before it looks like I'm just making things up.

If they exist, these cosmic strings are strange creatures indeed. Since they're generated when the GUT era ended—when the strong nuclear force split off from the unification—their exact width depends on the particular GUT theory you're paying attention to at the moment. But no matter what, they're no thicker than a single proton (because it's the strong force that determines the width of the proton).

Cosmic strings themselves have no mass. They're not *made* of any substance. But as a frozen-in flaw in spacetime, they have enormous amounts of tension. In these regions, spacetime is forced to have a fold it doesn't want to have, and is constantly trying to smooth itself out, but simply can't, like a stubborn wrinkle in your shirt that just won't iron out. In general relativity, any source of energy can affect spacetime around it: mass, energy, tension, you name it.

So despite their masslessness (is that a word? It is now), their tension means they act as if they had mass. So much so that an inch of cosmic string would outweigh a mountain, and a mile would outweigh the Earth.

And they're long. So, so long. These things are fundamental flaws in space, and space grows with time (we live in an expanding universe, remember), which means cosmic strings stretch with the universe itself. It's impossible to tell just how long they are, but a safe bet is "as wide as the observable universe."

And they vibrate, oh man do they vibrate. Kinks and cusps can zip up and down their lengths at the speed of light. (They're not made of anything so this isn't a big deal.)

Sometimes cosmic strings can twist and contort so much that they loop back in on themselves, or two cosmic strings can cross paths like giant lightsabers clashing. When this happens, a loop of cosmic string can pinch off, continuing to vibrate madly. These vibrations release gravitational waves (because all massive and moving things do, generally), and this release pulls energy out of the string, steadily shrinking it until it finally vanishes in a much more complex version of "poof."

An untied string can stay stable for eons, while some of the smaller loops might be evaporating right now, in the present epoch.

Strangely, you won't be particularly drawn to a cosmic string—a perfectly straight cosmic string has no gravitational attraction on surrounding matter (who knew that gravity could be so weird). But as soon as it starts wiggling, the game changes and you can certainly be pulled in their direction.

Since they're a fold in space, a circle drawn around them wouldn't add up to the usual 360 degrees. It seems completely nonsensical at first, but again, gravity is weird when we're willing to let gravity do whatever it wants. Imagine taking a piece of paper (I know, really stretching things here, but work with me). Imagine a circle perpendicular to that paper, with half above the paper and half below. Now fold a big crease into the paper. The circle will be lopsided, with one side slightly shorter than the other, thanks to the presence of that crease.

So if you find yourself going in circles that don't seem to quite add up, you just may have found yourself around a cosmic string. Try to give it a hug and you may find your arms wrapping around yourself.

If you have a headache by now, don't sweat it. We'll all in the same spaceship here.

It was once thought that they might be very common, and perhaps even responsible for the formation of the largest structures in the universe itself, with the filaments of galaxies anchored by the cosmos-spanning defects in spacetime. But studious observations have shown that they are—thankfully—elusive.

We haven't found evidence for a single cosmic string . . . anywhere, really. We've tried the age-old astronomy technique of looking for them with telescopes, and the relatively new technique of listening for their gravitational waves. It's all turned up nothing. Seems that our universe is surprisingly smooth and defect free.[11]

Which is actually a bit of a theoretical pickle, because cosmic strings are a natural prediction of GUT theories and our models of the early universe. And if we can't find any, then that tells us that either we don't understand GUTs, we don't understand the early universe, or we don't understand cosmic strings. Or we don't understand anything.

Cosmic string may still wander throughout the cosmos, and where they touch, they cause . . . well, let's just call it mayhem. Their thinness combined with their extreme gravity make them like the universe's most intense Ginsu knife. They slice. They dice. They never dull.

And, I should be specific, they can slice and dice *worlds*. Like butter. Sweet cream butter. There is almost no substance in the universe that can withstand their deadly combination of speed and density.

✪

The most surefire way to tell that you're near a cosmic string is notice that you're seeing double. Light from a distant object (say, a galaxy) can't go right through a cosmic string, so it has to take two paths, one on either side of the defect, so you'll get a split image of whatever you're trying to look at.

But once you get close to a cosmic string, it will be completely unmistakable.

Their extreme vibrations force kinks and cusps that saturate their nearby environs with intense gravitational wave fluctuations: imagine being relentlessly stretched in one direction and then the other, over and over until you're pulverized and/or ripped apart like a piece of putty.

We're not exactly sure how cosmic strings interact with the rest of known physics. It could be that they're completely silent and invisible, like a black hole only . . . longer, I guess. Silent, deadly knives forged in the earliest moments of the universe, vicious and relentless.

Or it could be that they're constantly churning and spewing high-energy radiation and cosmic rays, a vast, elongated source of pain, bright enough to be seen in the vastness of intergalactic space.

No matter what, they are perhaps the universe's most deadly and efficient killers. And they have friends.

Strings are the one-dimensional flaws in the universe, the king of defects, but imperfections come in many forms. Domain walls, textures, and *skyrmions* are all relatives of the cosmic string, differing in their number of dimensions and their configuration. Even our old friend, the magnetic monopole, may be just another type of defect. But they all manifest the same horrible property: absolute, implacable destructiveness.

✪

Wow, pretty intense stuff, if I do say so myself. Let's have a little bit of fun. Lighten the mood.

Imagine: you're skiing one day, enjoying the fresh alpine air and powdery snow. Down the hill you go, racing away. You come upon a slight rise, and not paying attention you trip over a rock, clumsily and embarrassingly stumbling to the ground. You're done, stopped in your tracks, prevented from falling (or, perhaps more accurately in this case, rolling) further down the hill by the little rise.

Having read up on your fancy physics jargon technobabble, you could remark to yourself they you are now in a *metastable* state. Not unstable: that's the falling part. And not entirely stable, either; that's the bottom of the mountain, waaay down there. You're not moving, and you're at *a* bottom, but a nudge or a decent shove would send you tumbling down to the true bottom. Metastable.

OK, so what. Only a problem if you're skiing. Or if you're the universe.

The universe has already gone through several phase transitions as it cooled, jumping from a high-energy state (the top of the hill) to lower and lower energy states (further down the hill). With each splitting of the forces, the universe crystallized to a new reality until it "landed" in the one familiar to us, with four forces of nature and a zoo of particles, everybody with their own set of masses and spins and all the other properties.

Since it looks like the universe hasn't gone through any radical phase transitions recently (where "recently" means 13 billion years), the fields that make up the forces and particles in our universe are probably at their most stable state. The very bottom of the hill. It seems as though the reality that we know and love is the true ground state configuration of the quantum fields that govern our universe.

Or not. "Not" is definitely an option.[12]

Maybe the universe is just metastable. At a temporary divot at the edge of the cliff. Maybe the early phase transitions led the universe to this particle configuration of fields and forces, and it stopped here because it got tripped up by a rock. No big deal. Everything's cool. We won't go tumbling down the rest of the hill.

Unless we get shoved. Then it's a one-way ticket straight to the bottom.

What could shove a universe? Or shove a piece of spacetime? Randomness can do the trick. I said earlier that a vacuum isn't really a vacuum. You probably skimmed over that part, so I'll repeat: an empty patch of space is really full of fields, and those fields can fizzle and sizzle. If they really are in a metastable state, then it just takes one random unlucky fluctuation to send these fields over the edge.

Let's go back to the ice in the freezer, because we don't have enough analogies going on at once. The ice gets its start in the liquid water at some point, somewhere. But where? The crystals that form the ice need a seed point, a *nucleation point*, some speck of dirt or scratch in the glass, something to grab onto and start into the crystal-making business.

Once a random jitter in spacetime serves as a nucleation point, the cosmic ice cube starts forming, a bubble expanding at the speed of light. What's behind that bubble? Why, nothing other than a completely different universe, that's all.

And by completely, I mean *completely*. Within that bubble the fields of our universe will find a new configuration. Remember what happens to the electroweak force when the energies drop too low? It completely disappears and is replaced with two new forces: the weak and the electromagnetic. If your life depended on the physics of the electroweak force, it's not going to be pleasant in the new regime.

If we truly are metastable, and a new stability emerges, there will be no more gravity or strong nuclear forces, no more electrons or neutrinos. All will be wiped away and replaced with something new. A new *reality*. Our current and familiar arrangement of particles and forces were born in the phase transitions in the early universe, when symmetries were broken, fracturing once-unified fields into disparate entities.

And thus will it be in the new universe.

There's nothing you can do to stop it. There's no warning. *If* that nucleation event happens, the expanding edge of the new reality bubble races forward at the speed of light through the old (i.e., our) universe. By the time you see it, it's already happened. Just one instant, you're relaxing in your easy chair on your planet or spaceship, sipping coffee, and the next, nothing. The forces that govern your body—indeed, the very subatomic particles that make up your body—will be ripped apart and rearranged. Life may be possible on the other side, but it won't be *your* life.

Is it possible? Certainly. The process could be happening in our universe right now, at this very instant, spreading like a cancer.

How likely is it? Well, the last time the universe went through this kind of subatomic furniture rearrangement was when the weak nuclear and electromagnetic forces split ways. I didn't mention this earlier, but have you ever heard of the Higgs boson? If you haven't, it's a cool fundamental particle, and its job is to be the sheriff of this splitting: The Higgs does the job of setting the differences between those once-joined forces.

Since the Higgs boson played such a major role in the last transition, its mass can tell us just how stable our universe currently is. And current estimates of the Higgs mass put us . . . right on the line between stable and unstable.

As in, metastable.

★

Wow, talk about nightmare fuel. Dense fragments of magnetic fields. Strings the length of the universe garroting worlds. The end of the universe itself in a bubble of reorganization. Scary stuff.

Pretty much, without question, some of the most dangerous and destructive denizens of the cosmos.

But I have to be honest with you, none of these have, well, actually been observed. They're only theoretical constructs at this point. So you can breathe a little easier in your travels.

The hunt is on, however, and you never know what uncharted expanse of the universe might host one of these creatures or be the tipping point for the creation of a new reality. The chances are incredibly low of you actually encountering one in your travels, but should you meet one of these dragons your fate is certainly sealed in fire.

Sleep tight, don't let the (ancient) bedbugs bite.

Dark Matter

Dark made darker
Smooth and velvety
Mostly sweet, a little bitter
I really miss chocolate.

—Rhyme of the Ancient Astronomer

Look out at the multitude of stars in our galaxy, just begging to be explored. Look beyond them to the hundreds of billions of galaxies in our universe. Go on, *look*. Who knows what secrets they hold? What adventures they have just waiting to be experienced. What mysteries of the universe, from the quantum to the cosmic, they can potentially unlock.

Like most things, the universe itself is much more than it seems at first glance. We're surrounded by light. The fierce nuclear engines burning away in the hearts of stars. The soft and effervescent nebulae, tracing their filigree patterns inside their galaxies. The rhythmic calls of the pulsars, the urgent songs of the blazars. The whispering microwave echo from the big bang itself. Our universe is awash in light.

But that same universe is almost entirely dark.

And not the kind of dark that you might lamely identify as a mere absence of light, like a shadow or the deep void of interstellar space. No, this is a different kind of dark. A special kind of dark. A dark so profound that *it doesn't even know what light is.*

It's downright spooky, if you ask me.

As you travel from star to star, or hop from galaxy to galaxy, you will be immersed in the Great Nothing, the gaps between all the interesting things in the universe. We've talked about the vacuum and its various fleeting and lonely denizens already, but there's one thing I deliberately left out. A new form of matter that is so crafty, so devious, so invisible that you don't even realize that it's pouring through you right now, a giant unfelt waterfall of matter.

We call it the dark matter.

We don't know what dark matter is—that's why it's generically called *dark matter*, instead of something cool sounding and specific like *neutralino* or *dilaton*.[1] When particle physicists cook up new ideas for fundamental particles in the universe, they get creative with their names. When astronomers cook up new ideas for fundamental particles in the universe, they get . . . dark.

Dark matter itself is thought to be rather benign. It is called "dark matter," after all, and not "dangerous matter" or "highly reactive matter" or "for the love of all things don't touch it matter." Nope, just "dark." Can't see it. Can't be seen. It's just there, hanging out, probably streaming through you right now, and you don't even care. But there's a chance—a slim one—that in certain conditions dark matter can be dangerous.

It's a long shot, I'll admit up front. The chance that dark matter can harm you in your travels is so low that it's barely worth mentioning. But I want to talk about dark matter for a few reasons: a) it makes up 85 percent of the mass of the universe, so what you thought you were exploring is really just a slim piece of all that's out there, b) it can sometimes interact with normal matter (e.g., you), and that gives it a big enough chance to be hazardous so I feel compelled to tell you about it, just in case, and c) I just want to talk about dark matter.

For the longest time Earth astronomers didn't even know that dark matter exists. Well, to give them credit, it is rather hard to find. You look up at the sky with your eyeballs and all you see is the hot glowy stuff. You know, stars, nebulas, galaxies. The places you want to visit. To the ancient astronomer that's basically all there was to the universe. And then you build impressive new telescopes, and all you see is even *more* hot glowy stuff. You go on to construct giant orbiting observatories that can see infrared or X-ray radiation, and what do you get? Hot stuff. Glowy stuff.

Planetary nebulae. Stars. Molecular clouds. Protoplanetary disks. More stars. The intracluster medium. The galactic halo. Even more stars. Splat out all over the place like a clumsy paint job. Sure, some of its dim, some of it's even hidden, but on average it's hot and it's glowy.

So you can forgive early astronomers for thinking they've got it all in the bag.

It wasn't until they started weighing galaxies that they noticed something was off.

You read me right: weighing galaxies. Why not?

"How much does that galaxy weigh?" is a fair question, right?

Sprinkled here and there I've already told you how much some stuff weighs, or more accurately how massive some things are. It's a legit scientific question.

You just need to find some really big scales, and when you do, the numbers come out way bigger than you anticipated.

⭐

We'll start with the easy way first, then work our way up.

Let's call it Method #1, and all we need to do is count.

For example, start with the stars. If you're just staring at a random lonesome star it's kind of hard to figure out how massive it is, but if you have two or more in a system, it's as easy as just staring.

Long ago, right when science was just starting to become hip and fashionable, Johannes Kepler figured out a key relationship between the speed an object orbits a central body (like, say, a planet orbiting the sun), the distance of that object, and the mass of the central body. He had no way of explaining his stunning result, since gravity wouldn't be invented by Isaac Newton for another hundred years, but nonetheless he figured it out.

We can play this game with the sun itself. You can calculate how long it takes the Earth to orbit the sun. Hint: it rhymes with "rear." With some trigonometry and a few rounds of careful observations (the kind astronomers excel at), you can figure out how far away the sun is. Dust off your abacus and you can crunch through Kepler's math and get the sun's mass.

Kepler's equation is just a special application of Newton's laws, so it works for anything, really, as long as one of the things is orbiting the other thing. See a binary star system? You can calculate the mass of the stars. Boom, easy.

For lonesome stars it's a little more complicated than that, but with enough binary systems we can start to build up a pretty solid relationship between how a star looks (color, brightness, and so on) and what its mass is. It takes a lot of digging and lots of careful annotations on tables, and of course it's not going to be perfect, but it gets the job done.

Once you have a catalog of stars and their masses you can take it one step further. Point your telescope at the nearest galaxy. How much does it weigh?

OK, here goes: we know the masses of various stars, and we know how bright those stars are. So if a galaxy is made of a whole bunch of stars, we can take all the light blasting out of the galaxy and divide by the light of a single star. That gives us the total number of stars in that galaxy. And if each star weighs the same on average . . . ta-da! The mass of the galaxy.

Yeah, I know, it's not that simple. Not all stars are the same size. Not all stars are the same brightness. OK, fine, we can adjust for that, using our fancy complicated tables and by examining the light from a galaxy in detail.

And aren't there gangs of gas clouds loitering around the galactic disks? There are, but we can estimate the masses of all that other junk, too. Usually we get this from the temperature. See a giant gas cloud that looks like it's been hanging around for basically forever? Chances are that it's been hanging around for basically forever. (I hope you remember this from our discussion on stellar nurseries, because the only clouds safe enough to visit are the ones that are complacent exactly like this.) That means it has to be in equilibrium, which means that the gravitational pull of its weight has to be balanced by the expansive tendencies of its own temperature and pressure. Measuring the light output gives us a sense of its temperature (more heat = more radiation), so we can work backward and calculate the mass.

We can survey the Milky Way and nearby galaxies in detail, looking for all the gas clouds that we can, and build up a proper census.

Plus we need to add some estimates for neutrinos, cosmic rays, planets, and other miscellaneous galactic riffraff.

You put it all in the hopper and get a good estimate of the total amount of stuff in a typical galaxy, basically based on counting all the hot glowing bits. The answer? Let's call it four. Four whats? It doesn't matter, just call it four.[2]

✪

So, job's done right? Pack up all our gear and call it a day.

Not so fast.

Scientists, being the careful, thoughtful people that they are, actually want to double-check their results. Crazy, I know, but just let them do their thing and it will all be OK.

Introducing Method #2.

Let's keep going with Kepler's ideas. Safe, reliable Kepler. What a pal. Let's watch stars orbit the centers of their galaxies. Or even watch stars orbit the centers of the Milky Way. It takes a while for a star to actually complete an orbit—a few hundred million whiles, actually—so we won't ever hope to watch the whole go-round in a single human lifetime.

But we can use the shifting of their light to calculate their velocities. Pick a few individual stars (or even groups of stars if your resolution isn't all that great). Carefully examine that light to look for the telltale signatures of known elements. Notice that those fingerprints are slightly out of place—either shifted blueward or redward from where they are on Earth.

Congratulations, you just measured a Doppler shift; the same Doppler shift that gives an ambulance its characteristic wail as it rushes by you can shift the wavelengths of light emitted by a star. If that star is moving away from you, the light gets redshifted. If moving toward, blueshifted. And the greater the shift the greater the speed.

Vera Rubin was the first astronomer to nail these kinds of measurements in another galaxy, and she was able to build up a profile of the speeds of stars at various distances from the galactic center.

From there you can bust out your trust Keplerian Calculator—if you know the speed of the orbit and the distance, you can calculate the total mass of all the things inside that orbit.

Redshift. Blueshift. Do a little dance. Calculate a little velocity. Measure the mass inside the orbit. What do we get? Twenty-seven![3]

Not four, which was the result we got from the method of counting hot, glowy bits. Just by measuring the speed of stars, there's more to a galaxy than meets an astronomical eye. It appears at first glance that there's a lot of *stuff* inside a galaxy that isn't all hot and glowy. These two methods for measuring the mass of a galaxy should agree, but surprise! They don't.

Hmmmm.

How about a tiebreaker?

Let's try Method #3 and look at something bigger, like a cluster of galaxies. Clusters of galaxies are, well, clusters . . . of galaxies. Buzzing beehives hosting a thousand individual galaxies or more, and those galaxies really are grooving, cruising around a hundred of kilometers per second (I feel compelled to note that there's a lot of empty space in the clusters, so the galaxies only very rarely intersect, which is a relief).

Let's assume—for the sake of argument—that the galaxy cluster has been around for a good long while. Like a few billion years awhile. I mean, why not? If they *haven't* been around for a long while, then we wouldn't see any. Because there wouldn't be any.

OK, galaxy cluster. Been around for a long time. Just like a star, there's a teeter-totter of forces. Gravity is trying to do its favorite thing and pull the cluster tighter together, but the motions of the galaxies are trying to rip the cluster apart. Since clusters appear stable, these competing forces must be in balance. Any faster and the cluster would've blown up long before we got a chance to observe it, and any more mass and it would've collapsed.

By measuring the average speed of the galaxies, we can calculate the total mass of the cluster. No Kepler, no counting. Just straight logic.

The first person to do this seriously was Fritz Zwicky in the 1930s. He's one of the few Earth scientists I'll mention by name, because his name is just too awesome to pass up on typing, so I'll do it again just for fun.

Fritz. Zwicky.

There. Mission accomplished.

As usual, we have two ways to measure the total mass of a galaxy cluster: just adding up all the gas and stars, and also using the motion of the galaxies. They should agree.

The adding-glowing-stuff method gives you the usual number, four. The temperature method gives you another number, twenty-seven.[4]

Whoops.

✪

OK, let's try this one more time. Maybe Kepler wasn't as cracked up as he thought he was—maybe we shouldn't rely on him so much. Let's go with someone *really* smart. Like, Einstein smart. How about Einstein?

Method #4: Heavy objects bend the path of light around them. When we look at something really big, like a heavy galaxy or even a cluster of galaxies, there will also be some light from even more distant background galaxies. As the light of those background galaxies passes through or near the massive object in the foreground, the immense gravitational weight will literally bend the path of light, shaping it just like a glass lens.

Massive objects are able to do this trick on the light through their intimate relationship with spacetime: mass and energy bend spacetime itself, and since light must race through spacetime to get to its destination, any hills or valleys or crinkles in spacetime will alter the course of the light.

The upshot: any background light will get bent. We see this all the time, too! Often enough to give it a name: Einstein rings and Einstein arcs. Look it up; cool stuff.

We can use the amount of bending to figure out the mass that's making the lens. And guess what we get? Your favorite number, twenty-seven![5]

And this number doesn't care about Kepler or motions or stability or any other arguments. It's just a straightforward application of Einstein's general theory of relativity, the same theory that predicted that even the sun should twitch distant starlight by just a hair, which was successfully measured to everybody's surprise (except Einstein).

Let's recap. Result based on adding up all the hot glowy stuff: four. *All other methods of estimating mass*: twenty-seven. All the hot glowy stuff only accounts for less than a quarter of the mass in big objects like galaxies and clusters.

This isn't a matter of theory versus observation, of astronomers assuming the universe is one way but nature turning out to be different. No, this is a matter of *observation* versus *observation*. You can't escape the data. You can't hide from the reality. Different techniques for measuring masses on massive scales are giving different results.

Something funky is up.

✪

Now I know what you're thinking: what's with all this finding out how fat galaxies are? Do they need to go on a diet or what? Where's the dangerous part? We're getting there, slow one, but to understand the danger you have to understand the physics. That's just the way the universe works.

Speaking of the way the universe works, when you're faced with a situation like this—when the stuff you can see doesn't jibe with their motions—you get two choices. Either there's more stuff in the universe, or there's something wrong with your understanding of physics.

That's what physics is, by the way, it's the relationship between stuff and their activity. I'm using the word *stuff* a lot because . . . well, I need to be as general as possible. Physics can describe the motions of planets, or balls of gas, or subatomic particles. If it's a thing, physics has it covered.

Earth astronomers have been in this particular pickle before, of motions of stuff not matching up expectations based on their mass. In those days, the "stuff" was planets and the expectations were based on Newton's law.

Scenario 1: The orbit of Mercury just didn't make sense. Instead of having a nice normal stable law-abiding elliptical orbit, it *precessed*: the point where Mercury reached its closest approach to the sun advanced inch by inch year after year, like a giant-sized spirograph. Newton's laws of gravity could solve a lot of problems, but not this one. At least, not completely. The gravity of the other planets in the solar system caused most of the precession of Mercury, but not all of it. A real stumper. It had been an issue for hundreds of years, but hey, we've got bigger problems, like the plague and the New World. No time for poor little Mercury.

So what was the solution? Maybe there's more stuff in the universe, like a new planet closer to the sun that's affecting the orbit of Mercury. Let's call it *Vulcan*, because that sounds totally awesome. Or maybe Newton's laws are wrong. Let's call it *Relativity*, because that's what Einstein called it.

The winner in this fight was the laws of physics. No new stuff in the universe, but updated laws. With Relativity we could perfectly explain the orbit of Mercury. Too bad, because Vulcan was a pretty cool name.[6]

Scenario 2: There was something wrong with Uranus. Stop giggling. Its orbit didn't make sense either—what we knew about gravity and all the invisible tuggings from the other planets in the solar system couldn't account for its motion. New laws of physics? Nope, just Neptune. A new planet. New stuff in the universe.[7]

Both solutions have worked before in our solar system: changing the laws of physics to agree with observations, and adding never-before-seen stuff to the universe. So who takes the cake with our latest observations of galaxies and galaxy clusters? New laws or new stuff?

Spoiler alert: new stuff.

In the later decades of the 20th century, there was a lot of debate, which is one of the favorite pastimes of Earthbound astronomers. Perhaps we were getting the laws of physics wrong. One of the leading contenders for new laws of physics, for modifying Newtonian dynamics, was called . . . modified Newtonian dynamics. This theory altered the way acceleration works in order to fit the observations of orbital velocities inside galaxies. OK, great.[8]

In this view, there's nothing hidden within a galaxy, we just happen to get gravity wrong on galactic scales—it turns out that Kepler wasn't up to the job, after all.

And we can go more advanced than these simple modifications to acceleration. After all, at the end of the day it's Einstein that rules the gravitational roost, so if you want to explain motion on cosmic scales then you have to unseat that current reigning champion, which means finding a replacement to general relativity.

Which physicists have gone and done about a hundred times and probably will continue to do so until time immemorial (or at least, until Relativity is finally and fatally unseated, if it comes to that). Modifications and extensions to Relativity crop up every few months or, all trying to skate the delicate line between being able to explain all the known physics (like exquisite tests performed on the surface of the Earth) and all the unknown physics (like dark matter).

All of them come up short. There are just so many different angles on dark matter that it's hard (or should I say, impossible) for a new theory of gravity to connect them all. It's like a game of scientific whack-a-mole: explain rotation curves, but you can't get gravitational lensing. Tweak Relativity to do away with the lensing problem, and you can't make galaxy motions work.

Physicists can be rather clever people, when they put their minds to it. But despite all that cleverness they just can't come up with a new theory of gravity that explains all the observations—the motions of galaxies, the velocities of stars, the lensing, and a lot more that I haven't even bothered talking about—at the same time. It doesn't mean such a theory doesn't exist (we're not *that* arrogant) it just means that we can't explain the data with that route. So given the choice between "adding new stuff to the universe" and "new theories of gravity that don't really work," we're going to have to pick door number one.

★

And what's behind that door?

"It doesn't matter," hear you saying. Oh no, it matters. It matters *a lot*. Because there's a lot of matter. Ah, I crack myself up.

Here's a picture of the universe that the observations are painting for us so far: most of the stuff in the universe isn't hot and glowy. It didn't have to be, did it? Why would we assume that everything in the universe lights up? How did that assumption get stuck in your brain in the first place? We already know about a whole bunch of things that don't light up: black holes, failed stars, etc., so why should "dark matter" come as any surprise?

Maybe that's the ticket: maybe most of the stuff in a galaxy is just normal everyday happy-go-lucky matter, but just . . . unlit. Not *dark* matter, but *dim* matter. Fair enough, we can run with that. Let me spitball another possibility: a new kind of fundamental particle that doesn't interact with light. Just tossing it out there; don't take it seriously. Yet.

We've got our four observations so far. That's led us to the conclusion that dark matter exists. But what *is* the dark matter, and how could it be dangerous?

This is the part where we get to use the whole entire universe as a fun physics experiment. You know the big bang, right? The history of our universe? On average, galaxies are getting farther away from each other (yes, that means that it's getting harder and harder every passing day to accomplish intergalactic adventures, but that's just our luck). This means that in the past our universe was smaller. And if it was smaller, it was hotter and denser.

Long ago, it was so hot and so dense and just so dang *crowded* that the entire universe was just a giant ball of plasma—the same kind of plasma we've seen inside the sun. Or a lightning bolt. Or those fun balls where you touch the glass and an arc tingles your finger.

Anyway, once the universe cooled off from *that* bit of fun, it released a bunch of radiation. At the time it was literally white-hot, but over time eons do what eons do best, which is to mellow things out. Nowadays that light remains, but it's shifted all the way down into the microwave regime, barely keeping its stuff together a bare few degrees above absolute zero—the cosmic microwave background. This background light is Not a Threat to you or your livelihood, and so I'm not going to talk about it much.

Astronomers and cosmologists have had a lot of fun over the decades studying this background, because it's like a giant dinosaur fossil up in the sky—it tells us what the universe was like billions of years ago. And more importantly, it tells us what the universe was made of. Radiation? Normal matter? Dark matter? How much of what components? What's the recipe here that best explains what we see in that ancient light?[9]

We can push things back even earlier.

In the 1940s physicists got pretty dang good at understanding high-powered nuclear reactions, and since then we've been able to build peaceful nuclear reactors and less-than-peaceful nuclear bombs. The extremely early universe—way earlier than the plasma party that led to the cosmic microwave background, I'm talking the first dozen minutes or so—was a feisty nuclear furnace, where the lightest elements were first forged. Using our nuclear-powered math, we can predict how much hydrogen, helium, and other fluff ought to be in the cosmos today.

Those predictions? The universe ought to be about three-quarters hydrogen, a quarter helium, and a tiny fraction *other* (or, if you remember your astronomy jargon, a "metal"). Which is basically exactly what we observe out there in the real universe.[10]

And the pictures painted by both the nuclear calculations and the cosmic microwave background sky are crystal clear: there simply isn't enough normal matter to account for the dark matter. For example, if you want the dark matter to be merely dim, then it has to be made out of the same stuff as you and me, just not very bright: planets, black holes, whatever. But planets and black holes and whatever must trace their origins to the thick and creamy primordial soup of the early days of the big bang, and all our observations and calculations and ruminations are telling us very clearly that there simply isn't enough raw material to make up the difference.

The amount of normal, light-loving matter in the universe is fixed at an almost insignificant 4 percent of the total amount of energy in the universe, an additional 23 percent being something else that's very, very dark. Now you know where I got those numbers from earlier.

(Aside for the curious explorer: the greatest source of energy in the universe, sitting at 73 percent, is called dark energy and is responsible for the nefarious act of accelerating the expansion of our universe. No, I did not make up that name on the spot and yes, astronomers really are getting that unimaginative with their nomenclature. However, for our exploratory purposes that is just a mere footnote and not of great or immediate concern to us, and so we will simply let it be.)

So all of our independent measurements limit the amount of normal matter in the universe. And our attempts to fix things by playing with our understanding of gravity are coming up laughably inadequate. Filling up galaxies with black holes, failed stars, and dim blobs of gas just isn't going to cut it, and neither is a new kind of gravity.

To explain what we're seeing, you can't just go dark, you have to go to *maximum* dark: a new kind of particle.

✪

Now I know what you're thinking: this is nuts. We're just going to go ahead and *invent* a new kind of fundamental particle just because the astronomers don't know how to count? At first blush it seems downright preposterous, but what exactly are our options now? You got any better ideas? No, I didn't think so.

Whatever this new *kind* of matter is, it completely unknown to previous human experience. And whatever it is, it can't interact with light, otherwise we would . . . well, we would see it, wouldn't we?

It's not as crazy as it sounds. Well, OK it's kind of crazy, but that doesn't make it wrong.[11]

We've made up particles before, even particles that don't interact with light. Remember the *neutrino*? No? Oh, well, don't worry too much, because they're not particularly dangerous, and therefore not a primary subject of this book. Sure, they're responsible for supernovae, some of the most powerful explosions in the universe, but you can hardly blame them for *that*.

Neutrinos were invented in the middle bits of the 20th century to explain some funky results in high-energy particle experiments.[12] And it turns out that they actually exist. And get this: you're *swimming* in neutrinos. Neutrinos pop out of nuclear reactions, and the sun is a giant blazing nuclear furnace: it's practically a neutrino factory. Hold up your thumb to the sun. It doesn't matter if it's nighttime: this still works, because the Earth is like a clear piece of glass to neutrinos. There are about 60 billion (yes, with a B) solar-made neutrinos passing through your thumbnail every second.

Try not to freak out thinking about it.

And before you jump up shouting "Eureka!," no, neutrinos are not the dark matter you're looking for. For a long time, we thought that they were completely massless, and it's hard for a massless particle to be responsible for the dark matter. But more recently we've learned that they do indeed have a tiny bit of mass (we're still not sure yet exactly how much, but that's not our immediate problem). For another, they're too hot.

At the biggest scales our universe looks like a giant, somewhat-creepy spiderweb, made of long, thin ropes of galaxies all tangled and knotted together. It's called the cosmic web for obvious results, and it's the end result of over 13 billion years of gravity trying to make things Bigger and Better. Yes, even our own Milky Way is a part of a large group of galaxies, unromantically called the Local Group. Our nearest neighbor is the Virgo Cluster, one of the same kinds of clusters of galaxies the old Fritz studied way back when.

Both the Local Group and the Virgo Cluster are part of an even bigger complex, the Laniakea Supercluster.

This supercluster hasn't always been here; it had to slowly get put together piece by piece. And we can simulate the evolution of structures like that on a computer. Physics is a story of math, and computers are really good at math. They can win on *Jeopardy!* and they can make pretend universes. And using the Power of Physics, we can change the ingredients of the universe and see what cooks up. Just like a cupcake.

A delicious, moist, buttercream-topped cupcake.

I give you a cupcake. This is a metaphor for the universe, OK? You want to know what the cupcake is made of. Change the ingredients in a cupcake (the stuff the universe is made of), change the recipe (the laws of physics), and you'll end up with different kinds of cupcakes, which you can compare to the cupcake I gave you. Preferentially through many delicious experiments.

When we bake our pretend cupcakes—I mean, universes—with different kinds of ingredients, we can compare them to maps we've made of the cosmic web. There are subtleties. There are always subtleties, but it works.

The problem with neutrinos is that the batter comes out too smooth. They're too light and too fast, and with too many neutrinos (like, enough to explain all the dark matter), normal matter has a hard time clumping itself together to build important things like galaxies.

In short: if neutrinos were the dark matter, galaxies wouldn't exist.[13]

To reproduce the cosmic web that we actually see, the dark matter has to be *cold*, which is a snappy way of saying *nonrelativistic*, which is a technical way of saying slow. Neutrinos are just too dang fast, and with too many of them they wash out anything smaller than a galaxy. Including you, and even worse, me.

The dark matter, whatever is, doesn't hinder the production of galaxies, but rather supports it. Every single galaxy that you see with your eye or optical amplifier of your choice represents just a small fraction of all the matter in the universe, and is just a small speck of brightness embedded inside a much larger, much darker, much more invisible *halo* of matter. A lighthouse on a distant, hidden shore. The galaxies of our universe are just so much flotsam and jetsam in frothing mess, one that we can barely even begin to discern.

⭐

So we know that the dark matter has to be some sort of particle, some kind previously unknown to particle physics. And we know it has to be abundant (by definition, because there's a lot of it), and it has to be heavy (again, otherwise we would've seen it by now in our experiments), and it has to be cold (to allow for the formation of structures that we know and love today).

We also know that this mystery particle (Or particles! But let's not get ahead of ourselves.) responds to the force of gravity—that's exactly how we spotted it in the first place. And we know that for all intents and purposes it is not on speaking terms with the electromagnetic force: otherwise we would see dimming or scattering or reflection or emission or absorption or any of the usual *astronomical* things that stuff in space tends to do.

What about the two nuclear forces, any game going on there? Well the strong nuclear force is out; it has its own set of particles and only operates at extremely short scales. But weak nuclear? Maybe there's something there.

We have other observations too that clue us in. Some of the best places to study dark matter are colliding clusters of galaxies. Massive train wrecks of gas, stars, and any other stuff. Like the wonderfully named Trainwreck Cluster.[14] When clusters collide, the gas gets all tangled up in the middle. Local TV meteorologists would go insane drawing all the fronts and storm systems. The individual galaxies comprising the clusters pretty much pass by each other, two clouds of flies. And the concentrations of dark matter, based on lensing measurements? Also completely pass by each other. No tangling, no communication, no interaction. Two ex-lovers passing on the street. Not even sharing an awkward glance.

Hmmm, that's an interesting clue indeed.

To make it all work, to make it all hang together, to match the observations of star velocities, of galaxy motions, of gravitational lensing, of light from the early universe, of the amounts of hydrogen and helium, and of the arrangement of galaxies in the cosmic web, we need dark matter to be a brand new kind of particle, that's cold and doesn't interact with light and is heavy enough to explain all the concentrations of mass that we see.

It also can't interact with *itself*, otherwise it would get all tangled up when clusters collide.

Our dark matter has to be made of a particle that is massive, but at best only weakly interacting with its own kind or any normal matter. A weakly interacting, massive particle. A WIMP, if you will. Physicists are so cute.

When I say, "dark matter," think "apathetic matter." WIMPs just don't care about you. Nothing personal, it's just the way the universe is wired.

Anyway, particle physicists are creative lads and ladies. They build crazy mad-scientist-style accelerators, bash atoms together, and get busy giving names to all the bits laying on the floor. When their accelerators are down for maintenance, they draw diagrams of absurd collisions on chalkboards to entertain themselves. "What if this quark slammed into a neutrino, but at the last moment a photon passed by?" (Just to be clear: that's a made-up scenario, but it's pretty typical of what you might overhear in a typical physics department hallway.)

With all their imagination and free time, they start cooking up ideas for particles. These new particles are by-products of new theories of fundamental interactions, forces, and laws. You speak into existence a new physical theory, and usually there's some extra baggage, like a new kind of particle. This is great news for physicists, because it gives them something to look for in their big colliders. You know, to test the theories and all. Science.

The theoretical particle physicists have been working diligently for the past few decades to move us past the Standard Model of physics, which is, as the name might suggest, the standard model of physics, which attempts to put all particles and forces and interactions under a single mathematical roof. Of special interest to those theorists is a particular high-energy range where the forces of nature just begin to unite (for the morbidly curious and those willing to go on a Wikipedia bender: the *electroweak* scale, which we met earlier when we studied in grim detail the looming disaster of a potential phase transition of the entire universe.).

It just so happens that a particle with that exact same energy range would fit all the properties we need to be the dark matter. The right mass, the right level of barely there interactions, the right abundance in the universe.

It's pure unadulterated coincidence—the theorists just happened to create the particle that the cosmologists were looking for. And if that particle is present and really does explain the dark matter, then it does indeed interact through more than gravity: it can talk to normal matter through the weak nuclear force.

Only occasionally, only fleetingly, but talk it does. Which makes it, potentially, detectable.

You can see why people got excited by a WIMP.

☉

Ah, the weak force. Doesn't get a lot of love, does it? Your life is almost entirely ruled by gravity and electromagnetism. And the strong nuclear force, while short ranged, is at least *strong* and holds atoms together. But the most charitable description of the weak force is that it plays a role in certain particle reactions. It's the wallflower of the fundamental forces. Always there, but not really saying anything. Not exactly a blast at parties.

But through this weak force the dark matter can potentially—maybe, possibly, perhaps—occasionally speak up and interact with normal matter. So with the billions upon billions of dark matter particles streaming through your body right now (just like the neutrinos), maybe every once in a rare while they acknowledge your existence. This means that we can build dark matter detectors and try to snatch a stray particle here or there. It will take a few years, but hey, that's job security.

To date nobody has snagged a WIMP. And to make matters a little more unsettling, the theories that lead to specific dark matter particle candidates (i.e., a mug shot for the actual WIMP suspect) are getting eaten away in our latest high-energy experiments. There's still a lot of room for a dark matter particle candidate, but it doesn't look like it's going to be something immediately obvious. When this matchup between the subatomic and cosmic realms was first uncovered, it was hailed as the "WIMP miracle," but it may not be so miraculous, after all.[15]

Still, we have our mountain of evidence that dark matter is a thing, we're just a bit lacking on the definition of the word "thing." And if dark matter does interact through the weak nuclear force, then it can be hazardous. Maybe.

The way the interactions happen—and how often they happen—naturally depends on the specific models, so I won't get into too many tangling details. The point is that rarely can a dark matter particle knock into a regular matter particle and cause a brief moment of subatomic panic. You don't really have to worry about that, ever.

But dark matter can also potentially talk to *itself*, which might be an issue. The dark matter largely ignores itself, but when it gets into dense enough situations two dark matter particles might find each other and go *poof* in a brief burst of high-intensity radiation.

Wait, wait, wait. If dark matter can annihilate itself, then why is there any dark matter around at all? Oh, you're a clever one, aren't you? It's not exactly common. It's *weakly* interacting, after all. You need millions of possible encounters—or even more—before you can even *think* of having or two dark matter particles interact.

But there are places in the universe where there's a whole heck of a lot of dark matter: the party zone itself, the galactic core. Dark matter makes up the bones of the cosmic web, after all, and the normal light-bearing matter flows to the highest concentrations of dark matter. So the hearts of galaxies also (we think) are home to abnormally high amounts of the dark stuff. Yes, I've already warned you off visiting the galactic core, because of its supermassive black hole and its intense radiation. Here's another reason not to visit.[16]

In fact, just to briefly dive into this dark rabbit hole for a bit, there might be *too much* dark matter in the cores of galaxies. Our fanciest computer simulations, which we use to trace the evolution of dark matter over the course of billions of years of cosmic gravity (it's just gravity and time, so not the most difficult calculation in the world) predict an absurdly large amount of dark matter in the cores. If they were correct, then essentially all the normal matter in the entire galaxy would be crunched up into a tight little ball.

This is most obviously not the case, so something's off. It could be that we don't fully understand the gravitational relationship between dark matter and regular matter (which wouldn't be all that surprising). Or it could be a hint that our naive models of dark matter are—Gasp!—slightly wrong. Various theorists have proposed solutions, and some of these solutions involve a new force of nature, one only felt by the dark matter and the dark matter alone, that allows it to have some extra interactions to smooth everything over and not get all hot and bothered in the cores.[17]

We don't know if there are extra dark-only forces of nature. We don't know if there's more than one "species" of dark matter responsible for what we observe. (The arguments here can go either way: either the largest component of our material universe should be Big and Dumb, or life in the dark sector should be just as crazy and expressive as our normal matter adventures, sorta like a dark chemistry and dark subatomic physics thing going on.)

We honestly don't know. We're kind of . . . wait for it . . . in the dark over here.

But we do think that when two dark matter particles interact, no matter what exactly they're made of, they can destroy themselves in a flash of gamma radiation. High-energy light. The hard stuff. Sure, one photon of gamma rays isn't going to harm you. You won't even notice a bunch of them. But *a lot* of gamma rays? Oh boy.

Gamma rays are *ionizing*, which means they rip electrons out of atoms and molecules. Big deal? Big deal. That means they can cause cellular damage. Tissue damage. Organ damage. *You* damage. In small quantities, you just get a slightly elevated risk of cancer. In large doses your guts melt.

There are searches underway right now for an excess of gamma rays at just the right energy range to match our WIMPy particle candidates. So far, nothing detectable above the general galactic gamma-ray goings-on, but we're still hunting.

Beyond that, it may be possible that dark matter can just out-and-out decay, randomly one day deciding that it's had enough of this galactic life of invisibility and just vanishing altogether, leaving behind a shower of more familiar, but more deadly, particles.

Moral of the story: don't go near galactic cores. There are other sources of gamma rays, for sure, but just an elevated background due to dark matter can cause your lymph nodes to start getting their own rebellious ideas. If you start feeling severe symptoms, such as headache, nausea, vomiting, diarrhea, cognitive impairment, decrease in white blood cell count, dizziness, hypotension, hypertension, tension, or death, you may be a victim of dark matter annihilation poisoning. Call a doctor immediately.

Who knew? Dark matter by itself is safe and relatively harmless. Well, less than friendly. It actively avoids you, which is even better. It just goes about building the universe as we see and love it. But get near a large enough clump of it, and there goes your lunch. You can't even see it coming, unless you have your gamma-ray detector activated and calibrated. You did have the thing inspected before you left, right?

Hostile Aliens

Are we alone?
Oh God, I hope so.
            —Rhyme of the Ancient Astronomer

For some reason, us humans are really invested in the idea of not being alone in the cosmos. Maybe it's the sweeping majesty of the heavens above, with all those glittering stars and effervescent nebulae, that calls to us, harkening to explore and expand and discover and learn, and perhaps most tantalizingly, to meet and communicate. It's all just so *romantic* isn't it? The thought that there might be other life out there, somewhere, in the vast expanses of the deep fills our hearts and imaginations.

What would they look like?

Could we talk to them?

Could we have an interstellar game night? Jenga or charades? Both? Both.

Or, alternative: those same vast expanses of the deep void do indeed call to us, but not with the warm whispering words of a mother calling to her children, but with the harsh foreboding tones of warning. Of fright. Of coldness.

Of loneliness.

It takes beams of light years to hop from star to star, millions more to hurtle between the galaxies. The Earth is one planet amongst eight (or nine, or ten thousand, depending on who you ask, so don't ask) planets orbiting the sun. The sun is one star amongst three hundred billion (give or take a hundred billion; we're still counting) making lazy circles around the center of the Milky Way galaxy. Our home galaxy is one amongst . . . uh, a lot of galaxies, somewhere between 500 million and 2 trillion (look, it's really hard to get an accurate headcount when most of the heads you're counting are billions of light-years away and also very, very dim) within the observable universe. The observable universe is our local patch of the party, about 90 billion light-years across, and just one small blob of an otherwise much, much larger cosmos. How large? Estimates here range anywhere from 10^{62} times bigger than our observable bubble to *literally infinitely big*.[1]

The total amount of stars in our universe is some very large number that I'm not even going to type out because it would be completely meaningless and simply serve to entertain you with the number of zeros I managed to type out. And that doesn't even count the nebulae, degenerate leftovers, planets, planetesimals, dwarf planets, brown dwarfs, asteroids, comets, meteoroids, cosmic rays, neutrinos, radiation, and on and on and on.

That's a lot of stuff, and that's a lot of volume.

Our universe is too big to describe; too big to even begin to comprehend. Our evolutionary heritage just didn't equip our poor little lizard brains to handle such concepts as the bigness of The Bigness. Thankfully, we invented mathematics, which like any tool allows us to do things we weren't normally meant to do. Good luck sawing down an oak tree with your bare hands, and good luck trying to encapsulate the expanse of the cosmos by just thinking about it.

Face it: the universe is big. And that largeness is, in no small way, absolutely terrifying. To think that here we are, four-limbed ape descendants who for all practical purposes just gave up life in the branches *yesterday* are the sole inheritors of the supreme majesty of all creation is . . . a bit hard to swallow, don't you think?

Or maybe you don't think about it, because even thinking *about* the thought is too terrifying.

The same night sky that calls and beckons us to explore it also manages to crush us with its incomprehensible depth and breadth. The universe *exists*, far larger than we're ever comfortable imagining. And it's all for us?

Perhaps we desperately want intelligent life to be out there—and search in vain for it again and again—because we can't really fathom trying to do all this on our own.

Will somebody please just pick up the phone?

★

Ancient philosophers and early astronomers imagined a multitude of worlds with life-forms aplenty, because philosophers and early astronomers had a lot of time on their hands (apparently) and enjoyed cooking up random ideas without any grounding in reality or evidence (apparently). It's easy to pretend that there's a lot of intelligent species zooming around the universe; it's quite another to actually find evidence for them.

Just look at the whole funny business with the Martian "canals" in the late 1800s. Some seriously professional and professionally serious astronomers looked at the Red Planet through their then-powerful telescopes. Understandably, they saw a bunch of vague fuzzy blotches and features. One Italian astronomer, Angelo Secchi, in 1858 missed a hilarious opportunity to call some of these features "cannoli" and instead called them "canale." This Italian word looks like "canal", and can translate into that if you felt like it, but also into something much vaguer, like "gully."

Either way, astronomers through the next decades outdid themselves in depicting what they thought they were seeing through their eyepieces: complicated canal networks, evidence for forests, even potential engineering projects actively underway. Why, it was just like civilization on the Earth, but redder. Neat!

Not every astronomer went Marsballs, and were quick to point out the naive and fanciful idiocy on display by their colleagues.[2] By the time the early 20th century rolled around, improved telescopes and a fancy new technology called *the camera* put to rest the idea of Martian-made artificial features. Turns out it was just a bunch of boring red sand.

Now I'm not going to sit here and tongue wag at those hapless astronomers (too much), except to say that they were really, really desperate for companionship and would've loved to have some friends on our neighboring planet. Whether they were driven by the urge to form common bonds across the depths of space, or motivated by the fear of the lonely void, it doesn't matter—they, like us, wanted some buddies.

The whole will-you-be-my-neighbor schtick didn't end once we realized that Mars is both red and dead. Throughout the decades astronomers here and there have thought they caught the whiff of an alien scent, furtively couching their results in sufficient scientific jargon and caveats to make sure none of their peers dismisses them outright for being, you know, a crazy, but slipping in just enough language to get some attention, and should their result be the first true indication of an intelligent species outside the Earth, instant and total fame and possibly fortune (at the very least, a memoir and a biopic, and maybe even some merchandising opportunities). Oh, and a Nobel Prize, I guess.

Do you remember pulsars? I hope you do, because you're never, ever supposed to get close to one, unless you want your face to be melted off from the intense X-ray radiation, followed shortly by your entire body dissolving from the unholy magnetic fields. Good, never hurts to have a friendly reminder.

Pulsars were first discovered in 1967 with a radio telescope, accidentally. Antony Hewish and his grad student (now Dame) Jocelyn Bell Burnell were monkeying around with the Interplanetary Scintillation Array, which sounds way more awesome than it really is, which is a bunch of wires hanging off some poles.

Anyway, one night they found a radio source that repeated itself precisely every 1.33 seconds, on the dot, from the same location on the sky, on the dot. What could it be?

It wasn't a star—stars don't act like that. It wasn't a galaxy—galaxies don't act like that. What could possibly make something so regular and monotonous? Could it . . . be? Burnell and Hewish "jokingly" (their word) referred to the source as LGM-1: Little Green Men.[3]

Soon after, another source was discovered, and another, and another, and another. The sky was littered with LGMs! We're about to be invaded!

Or, as astrophysicists were quick to point out, it was the beam of a rapidly rotating neutron star washing over the Earth like the blinking of a distant lighthouse. Just another random exotic thing in the universe. Perfectly natural and definitely not aliens.

We ended up calling them pulsars instead of LGMs. Fair.

That wasn't the end of little bleeps and bloops in the sky being mistaken for alien calling cards. In 1977, the Big Ear radio telescope (Earth astronomers have a certain flair for names, especially when it comes to their pet instruments) outside Columbus, Ohio, recorded an exceptionally strong signal lasting seventy-two seconds. The radio flash was thirty times stronger than the background radio fuzz of the universe. It also occurred at a very peculiar frequency: the frequency naturally emitted by neutral hydrogen. Coincidence?

The dude in charge of observing that night, Jerry R. Ehman, was suitably impressed enough to write "Wow!" on the printing. Hence the name that went down in the history books: the *Wow! signal*.

What was this signal? No clue.[4] It never happened again, ever, in the history of radio observations since that one lonely night in 1977, even with bigger and badder ears at our disposal. If it was aliens, it was a one and done, a quick "Hey" before shutting off their civilization for good. Or maybe it was something terrestrial (the military is understandably a tad hush-hush about their strong radio emissions), or reflections off a comet. Unless we get a second dose of Wow!, we'll basically never know.

The hopeful wistfulness of our search for aliens got more eager/sad as time went on.

Since 1998 the Parkes Observatory, a radio telescope in Australia, would occasionally pick up faint whiffs of mysterious radio transmissions (see a common theme here?) that would disappear as soon as they arrived, flirting from one frequency to another. The signal didn't appear to come from any particular direction on the sky, but did tend to cluster around midday. But no, it wasn't the sun. They checked. They were dubbed Perytons, because that sounded fanciful enough. Papers were written about them. Could it be . . . ?

But they didn't check their cafeteria. It turns out that when you get a little too eager for your food to come out of the visitor center's microwave, and yank the door open before the timer beeped, it took an extra fraction of a second for the stuff that makes the microwaves to actually shut off, leaking a little bit of radiation. And when you have a 64-meter radio telescope *literally next door*, you can trigger a flash of "mysterious" radio emission.[5]

People don't write papers about Perytons anymore.

In 2015 astronomers using the Kepler Space Telescope found one star to dim very oddly: big dips in brightness at random times, which is definitely something that stars shouldn't do when left to their own devices. Strange, right? Could it be . . . ?

Not a message, in this case, but perhaps, maybe . . . there are of course many potential natural explanations, but I'll wink every time I say, "we don't know." Maybe some intelligent critters built a bunch of weird giant structures around that star, and those giant structures are blocking the light when they pass in front of us.

I'm not saying it's aliens, but here's my contact info for media inquiries.

A couple years later more observations revealed the culprit: dust. Just a bunch of stinking dust spotted. Apparently, this star forgot to clean up its room.[6]

Later, in 2017, astronomers confirmed the first interstellar interloper to pass through our solar system, which was capable of blocking enough light to account for the peculiar dimming. A confirmed visitor, named 'Oumuamua (a Hawaiian word roughly translating to "scout"), hailing from some unknown origins in the deepness well outside our neighborhood, spending only a few short years looping around our sun before heading off to parts even more unknown.

It was a big rock.

An exciting rock, to be sure (and astronomers are ones to get *very* excited about rocks in space), but just a rock.[7]

Speaking of rocks, sometimes something big crashes into a planet, sending chunks of that planet flinging far and away, where they float around helplessly for a few million or billion years. Most of those chunks end up doing that and only that for the rest of their lives, but sometimes a few of them get swept up in the gravity of another planet. If they survive the descent, they can then sit in comfortable retirement in their newfound home.

Unless those chunks hit the Earth and are unlucky enough to be scooped up as a meteorite, where they will have to endure years of uncomfortable poking and prodding from lab-coated scientists trying to understand it.

One such rock, ALH84001 and *not* nicknamed "Alphie," is a four-billion-year-old rock on Earth, but made on Mars. Inside this rock are teensy-tiny little structures that could, if you squinted hard enough and made yourself believe it, be made from some long-dead microscopic organism. Maybe. There is also a raft of chemical or mineral (i.e., "boring") processes that could make the same features.

For now, it's just a rock.[8]

★

OK, so there's no evidence right here at our doorstep for any life outside the Earth, and none of the big flashy candidates have come up aces. Is there any sign, anywhere, of life or intelligence? (Let's assume for the sake of argument that humans do indeed score well on the galactic scale of "intelligence.")

This is where SETI comes in. Reader, meet SETI. SETI, reader. These are the folks Searching for Extraterrestrial Intelligence. I suppose they could have broadened their search and looked for any kind of extraterrestrial life, not just the intelligent kind, but that means they would've had to . . . wait for it . . . SETL.

Over the decades since the 1960s, astronomers—especially radio astronomers, since the presumption is that alien intelligences will be just as fond of tuning in to their favorite oldies station as we are—have taken slices of time from observatories around the world in their hunt for the unmistakably alien, "Hey how are you." It's understandably hard to dedicate an entire observing facility with this one goal in mind, because the universe is full of radio emissions of all sorts, and that data is full of juicy astrophysical information just waiting to be heard.

Sure, we're curious if we're alone or not in the cosmos. We just don't want to spend any money on it.

I won't let you wait around in suspense any longer than I strictly have to. Besides that one *Wow!* in 1977 (I'm talking about the mysterious radio signal, not the release of *Star Wars*), SETI hasn't heard anything. Zero. Zilch. Nada.[9]

Well, I mean, it's heard all sorts of things, just not an obvious calling card from our next-door (galactically speaking) neighbor. Which, once you start to put some numbers on it, doesn't seem so strange.

Yes, we have huge radio receivers capable of detecting the faintest whiffs of electromagnetic essence from across the universe. But the stuff we hear with our dishes is loud. Not loud-movie loud. L-O-U-D. We're talking about exploding supernovae, or rotating neutron stars, or swirling clouds of gas around giant black holes. The big stuff, the real stuff, the hard stuff.

Even if an alien civilization a) knew that we were here, b) decided to talk to us, c) built a giant radio transmitter, and d) blasted us with the biggest HELLO they could possibly achieve, our current technology limits us to a radius of about 100 light-years.

Considering that the Milky Way galaxy is 100,000 light-years across, our listenable bubble is pathetically tiny. It might as well be zero, *and* that assumes that the aliens are actively trying to talk to us.

Of course, future radio telescopes and arrays will be able to pick up fainter and fainter signals, which should in principle increase our range, but the further we go, the more we run into another issue: the galaxy itself. All the indistinct astrophysical processes in the galaxy (and heck, the universe) meld together into a general background radio hum, and if you want to pick out an alien signal (assuming, again, that aliens are trying to make a detectable signal in the first place), you have to dig that diamond out of all the rough.

It's not impossible, but I wish you good luck, because you'll need it.

Still, calling Sol isn't the only thing a supposed intelligent alien civilization can do. (Assuming they even *want* to do it. I mean, no offense, but if you had an unimaginable level of technological sophistication, would you really want to chat with . . . us?) They can also muck about with their environments.

You know, like we are messing around with planet Earth, with all our (actual) canals and parking lots and pollution and artificial islands. Heck, we've even changed our own atmosphere to a respectfully detectable degree (uh, we can just pretend that signalling our existence to the wider galactic community was the entire point of all our carbon emissions).

An alien intelligence can do whatever they want, if they've got the energy and free time to do it, I suppose. Perhaps they'll get hungry for power (as in, energy sources, not world domination, though it might entail both) and want to capture all the output from their stars. Lots of radiation just floats off into space being totally useless except to distant astronomers, after all.

So maybe they'll disassemble all their planets and build giant orbiting solar arrays, taking all that energy to . . . uh, well, who knows, but presumably they'd find a use for all that power.[10]

Of course the concept of these *megastructures* relies on some tricky math. To build the megastructure you need a lot of raw material. Most of the stuff in any given solar system will be hydrogen and helium, which are both next to useless. You need lots of rock, and most of the rocky bits are crammed down in the centers of gas giants. So you have to dig them out, either by diving down or by disassociating the planets themselves. And then you have to move all that material into the right orbits. And manufacture the solar panels themselves.

All this reconfiguring of a solar system costs energy, which you'd have to get from somewhere. You could bootstrap yourself and store up a little solar energy at a time to get on to the next phase of your interplanetary Lego project, or you could rummage around the basement for some nuclear material to really speed things up.

But no matter what, it will take time and energy (loads of both) to make a megastructure. Even up to millions of years. And once you start collecting all that juicy solar radiation, you might have to wait thousands or even millions of years more to recoup your initial energy investment. In most scenarios, you're better off just leaving things as they are. Nature put them there; don't mess with them.

Still, it's not technically impossible, so we might as well go looking for it.

If you completely enclose a star in a giant Dyson sphere (named after Freeman Dyson, the first person to propose a megastructure like this, but don't worry too much about it, since these spheres are unstable and will instantly crack open the minute you put the finishing paint job on it), it won't make the star go invisible. No solar panel and no way of extracting and using energy is 100 percent efficient, because physics. Some of the radiation from the star will still make it out, but way down in the infrared, as waste heat from the outside edge of the shell.

So presumably we could search for funky-looking stars that are large and mostly infrared. Heck, if a really eager alien civilization managed to cover up a good portion of the stars in their galaxy (hey, if you're gonna do it once, you might as well show off), then it could skew the entire light output from the galaxy itself, making it detectable even from the other side of the universe.

Besides that one oh-right-it-was-actually-dust example with the Kepler Space Telescope, we haven't seen a single megastructure in action in our galaxy or any other. Period.

When it comes to possible evidence of life outside the Earth, that's all we've got:

1. bleeps,
2. bloops, and
3. rocks.

Chances are that in your lifetime there's going to be a few more Big Reveals, with some group somewhere excitedly claiming that this time we did it, we really did it, we found (possible, maybe, somewhat sketchy) evidence for life outside the Earth. Maybe it will be an unexplainable radio signal. Maybe you'll see some ultrazoom-ins on some piece of space rock. Maybe the scientists will be dead serious, maybe they'll explain their results with a shrug.

But it's bound to happen, and when it does, you should have one reaction and one reaction only: skepticism. Supreme, ultimate, top-shelf skepticism. The kind of skepticism that makes you start to question your own existence as a thinking being.

Why? Because finding aliens would be a *really big deal*, and so far we haven't found any evidence of them, which means there's a solid chance we are (effectively, at least) alone. More on that in a bit. The upshot is that a claim that aliens exist and are really real requires a healthy slathering of evidence.

Positive evidence. Not "evidence" in the form of something unexplainable. If it's unexplainable, then by golly, you haven't explained it, with aliens or anything else. For example, the Wow! signal. We don't know what caused that mysterious radio burst way back in 1977, and that's the sum total of what we can say about our conclusions.

If you receive a random radio signal in your backyard receiver or see something funky going on with a nearby star, it's overwhelmingly more likely to be natural. And get this: your best guess is always to assume it's natural, even if aliens *really are the culprit*. Anything that can possibly have a natural explanation—in other words, if there's even a slim chance that nature herself just cooked up the signal from some exotic new way of blowing up a star, then that's the route you have to stay on. This is the way evidence works: you want to make a big claim, then you need a mountain of evidence. Period.

"It might be aliens" isn't good enough, because aliens, especially of the intelligent variety, are capable of creating any possible signal they want. See a big spike in a particularly interesting radio frequency lasting seventy-two seconds? Aliens *could* do it. No, wait, it was actually only half as strong, at a very uninteresting frequency, and lasted thirteen seconds? Well, aliens *could* do that, too. They can do whatever they want—they're smart cookies.

See a star dimming in an odd way? Well it *could* be a megastructure orbiting that star. But what if the dimming pattern changes, or was altogether different? Well then, no big deal for an alien; they can change their megatoy on a whim. They're aliens. They do what they want.

Not only does the cry of "aliens" require a major effort to hold up, but the *intelligence* in aliens make them an entirely unsuitable scientific hypothesis. A smart, clever, resourceful enough alien civilization can do whatever they feel like on any given day; their abilities are only limited by our human imagination. So they have the magical power to produce *any* kind of funky signal we get or observation we make.

Intelligent aliens are capable of explaining any piece of data. It's a hypothesis with maximum flexibility, which makes them useless (scientifically, of course, I'm sure they would be great workers, friends, and conversationalists). You can't actually test for this lazy hypothesis, because they can outwit your test by just being intelligent and having the resources to modify their environment.

Sure, aliens might be out there. They might be sending us signals. They might be messing around with a few stars here and there. But right now, there's no need to believe in them.

★

But are we really all by our lonesome, out here on some average planet orbiting an average star, sitting on the spur of an average spiral arm in an average galaxy? There are two, and only two, potential answers you can give to this question. Pick one:

1. We have absolutely no evidence—not a single solitary shred—of any life, as brainy as us or as dumb as a bacterium, anywhere else in the universe but right here on dear old Earth. And until we have positive evidence to point to (and preferably some sort of limb to shake hands/flippers/tentacles/pseudopods with), then we must assume that we are alone. Evidence is evidence and rules are rules. There's nothing to believe in until there's something to believe in.

2. I mean, come on, we've barely got started. Just *look* at all those stars, man! There are so many stars in the observable universe that I refuse to even type out the approximate number. Are the odds of life really so razor slim that we are the one and only place *in the entire flipping universe* to evolve creatures that can ask these sorts of questions? If—and that's a big if—nature is going to go through all the trouble of making life rare, then why didn't she go all the way and just make it impossible? Can you even point to any other object that only appears once in the entire history and volume of the cosmos? *I didn't think so, smart guy.*

It's a tough call. It's true that until we have evidence, we can't say anything conclusive. For now we're just stuck with thinking really hard and arguing even harder.

But the fact that both answers are plausible is itself an unsettling fact, if you're prone to bouts of unsettlement. The big trouble comes in our supreme *averageness* on scales galactic and cosmologic that I mentioned earlier. We're are so captivatingly boring, so much so that we've enshrined it as a base assumption of our models for how the universe works. We're not at the center, or the edge, of anything in particular. We're just here, hanging out, minding our own business, being ourselves.

There are a multitude of stars just like our sun. Presumably, there are a multitude of worlds just like the Earth. So, can you follow me here, where are the multitude of aliens like us? Where is everybody?

If intelligent life happened here, and we're just plain average, then the galaxy—let alone the universe—should be *infested* with critters all over the dang place. We should be hearing shouts from civilizations near and far, and see evidence for microorganisms in all manner of spacey locations.

But we don't. Silence. Dead, flat, foreboding silence.

So maybe we . . . are . . . special? Maybe there is something unique about us? Something rare? But that flies in the face of everything *else* about our particular stellar neighborhood. All signs point to "not different," and yet . . . it seems we are different.

This is the heart of the so-called Fermi paradox: if life happened here, once, it should happen all over the place, many times. But it didn't. So what gives?[11]

Honestly, we don't know. The chances of life appearing on any one planet are greater than 0 percent (otherwise you and me wouldn't even be here to have this conversation), and less than 100 percent (otherwise we would've seen alien critters around every single star, and probably started importing some of their TV shows by now). But what's the actual number? Are we alone in the galaxy? Are there a dozen other alien civilizations surrounding us? A thousand? A dozen thousand? More? Less? I'm spitballing here; somebody help me out.

There have been, of course, heroic attempts to "estimate" the number of intelligent civilizations in the galaxy. I put scare quotes around estimate for several reasons, the primary one being *that we have absolutely no clue, and we're all making it up here*. Tools like the Drake equation (which you may have already heard of. If you haven't, here's the skinny: it's an equation invented by astronomer Frank Drake that tries to estimate the number of intelligent critters out there) purport to break this very complex and large question into a bunch of smaller, more digestible chunks. Instead of trying to directly answer the Big One ("Are we alone?") it tries to answer a bunch of little ones.

Those little ones range from the somewhat reasonable ("How many planets are there in the galaxy?") to the just-as-impossible ("What are the chances of life, once it appears on a planet, to start making radio signals?"). Unfortunately for the Drake equation, just because the questions are smaller in scope doesn't make them any easier to answer. What people end up doing when applying the Drake equation is just parametrizing their own ignorance—put a bunch of guesses into the equation, out pops . . . a guess. You could've just started and ended with that.[12]

Maybe in your adventures you'll be the first to discover signs of life, intelligent or not, past or present. Some folks back on Earth have even started *broadcasting* our details, instead of just letting our radio emissions accidentally leak out into space, hoping that someone picks up on the other end. Some are optimistic, hoping that we can kick-start a new round of interstellar cultural exchange ("I wonder if we can eat their cheese," they ponder excitedly), while others totally freak out at the thought: if they can find us, they can . . . exterminate us? Colonize us? Pillage us?

The truth is we don't have much to fear from any other intelligent life in the galaxy. For one, they are so stupidly far away it's not even funny (quit laughing). For another, there is absolutely nothing valuable on the Earth (Water? Gold? Beachfront property?) that you can't find on a bajillion (estimated) planets all over the galaxy, including the asteroids, which are *so* much easier to get ahold of, because you don't have to fight with big gravity wells to get at your precious resources.

I know your mom thinks you're special, but humanity and the Earth as a whole don't really have a lot to offer the inquisitive and exploratory alien species.

Did I mention that we're far away? More on that in a bit.

Besides, we don't have to worry about broadcasting our existence to the greater galactic community.

They already know we're here.

✪

Let me tell you how to really look for aliens, on both ends of the scope, from the micro to the macro. And assuming that intelligence is intelligence across the universe, how they can look for us, too.

The first thing you have to hunt for is water. Look, I'm not going to sit here and pretend that life in the absolute general sense must require water, because life in the absolute general sense may not require water. We're not even exactly sure—and we've been thinking about this for a good long while—what life really needs to be life. The trouble is that we only have one single lonesome example of life in the universe, and you're a representative of that highly exclusive VIP club.

We know what *our* kind of life looks like, and assuming (for now) that there is some other form of life out there in the vast and frightening universe, we don't know if we're a pretty typical, average, mundane example (Sigh, *another* carbon-based terrestrial life-form? How basic.), or if we're the galactic oddballs.

We know that life doesn't require sunlight as an energy source. The vast amount of life on Earth needs the sun, for some reason or another (yes, when you eat your breakfast cereal you're really crunching on concentrated and repurposed star power), but some life found a comfortable niche around hydrothermal deep-sea vents. The deep sea doesn't get much (as in, zero) sunlight, but those critters (and they really are advanced enough to be called "critters") seem to be having a good time, taking energy from the complex chemical soup belching forth from the bowels of the Earth.

But from what we know, which isn't a lot but it's all we get, is that water *seems* to be a solid bet when it comes to life. At least, the liquid kind. Solid and gas forms of water just don't have the same flexibility (or too much of it). For an embarrassingly long time, Earth astronomers thought that the only place in the solar system with liquid water was dear old Earth, what with our oceans, streams, rivers, springs, and puddles. Don't forget the puddles.

They were wrong. Certainly Mars and almost certainly Venus had liquid water back in the good old days. Venus has a little bit of the H2O floating around the atmosphere, but good luck putting that to any good life-bearing use. And Mars? Well, Mars is tricky. It still has a lot of frozen water (it's far away enough from the sun for that, at least). Every once in a while, astronomers and planetary scientists (the people who want to study the Earth but have their heads in the stars) think they find slim evidence for some good liquid in that red desert. But then the evidence, like the supposed water, vanishes into thin air.[13]

There's also this question on Mars about the methane in the atmosphere, which seems to change up and down with the seasons. Methane is produced by microscopic bug farts on Earth, and microscopic bugs tend to thrive in warmer weather. So, mayyyyyybe there's something there? Maybe not—it's hard to tell without more digging.

All in all it's a tough call when it comes to Mars. For now, we'll leave the status of liquid water on that world as a nice and solid "maybe," and life there as a firm "to be determined."

But the outer worlds, man, they've got some surprises up their icy sleeves. Not the four giant planets themselves—their atmospheres are too thick, too dense, and too *weird* (as in, quantum pressure weird, but that's not important unless you're looking for particularly gaseous place to meet your final end) to host liquid water. But the moons of those planets tell a different story.

Yes, most of the moons are just boring lifeless chunks of rock. But some of those moons are boring lifeless chunks of ice.

On their surfaces.

And those surfaces are hiding some secrets.

It turns out that some of the moons aren't frozen all the way through; they're merely encased in ice. Their hearts are kept warm through very special gravitational relationships with their parent planets and the other moons. To be a little bit more specific because I know you're curious, some of their orbits aren't exactly circular, due to slight but incessant gravitational pulls from the other moons, like a kid sibling who keeps tugging at the bottom of your shirt. As the moons orbit around their parent, sometimes they're further and sometimes they're closer, which changes the gravity that they experience, and especially changes the *direction* of the gravity that they experience.

In the course of a single orbit, some moons will get stretched and squeezed like a piece of planetary silly putty. That internal friction is quite irritating, driving up the temperatures in the core high enough to melt rock. It's quite a different process than for the Earth (for our home world, the core retained some heat from its formation, and also keeps its fires lit through the decay of all its radioactive elements), but the result is the same: a mushy, squishy, molten core.

A hot core.

But the moons are way too far away from the sun to be warm on the surface, hence the outer crusts of ice.

Europa, Enceladus, Ganymede. Maybe even Pluto. Homes to liquid water—in most cases, more liquid water than the Earth has. Let that one sink in for a good long while.[14]

These oceans never, ever see the sun, but they're there. Could they host life? Maybe, maybe not. If those oceans come into regular contact with, say, hydrothermal vents, they might have that right magical mix of chemistry to turn a bunch of molecules into a bunch of self-replicating molecules that start eating each other. Are there giant space whales swimming through the oceans of Europa, buried under a hundred kilometers of rock-hard ice? I sure hope so, because that would be awesome.

And then there's Titan.

The largest moon of Saturn, it boasts an atmosphere thicker than Earth's, second only to Venus when it comes to atmospheric bulkiness around a rocky world. And Titan has lakes, rivers, streams, the whole deal. Oh, right, but not of water. Of methane and other hydrocarbons. At a temperature of -179 degrees Celsius (or in Fahrenheit, "really, really cold"). Too cold for water but the methane seems quite happy.

Could methane replace water's role in life on that strange world? Could *chemistry* find the right *bio* in that kind of environment? Are there space slugs oozing around in slow-motion at the edge of one of Titan's seas? I hope so, because that would be really awesome.

Did I mention that Titan is also probably home to a subsurface liquid *water* ocean? Yeah. Titan.[15]

Our own backyard is full of potential (emphasis on the *potential*) homes for life, simply because liquid water is ridiculously more common than we ever thought. Soon enough we may have to treat those worlds with an interplanetary version of, "Won't you be my neighbor?" Or not, we're kinda shooting in the dark here.

But why stop our searches in our own solar system, when there's an entire galaxy to explore?

★

And explore we will. And are literally doing it right now. And we don't even have to get out of our pajamas to do it. Indeed, this kind of exploration is best done in pajamas. I'm sure we can all get onboard with that, right?[16]

So here's what you do to hunt for signs of extraterrestrial life outside the solar system. Step one: get a telescope, the bigger the better. Step two: stare for a really long time at a random star. The more the merrier. Step three: look for any dimming from the star. Step four: be sure that the dimming isn't just caused by the fact that stars sometimes randomly dim for reasons that they would prefer to keep private. Step five: name your newfound exoplanet after yourself. Repeat.

Planets orbit stars. Look at the stars in the night sky; you know one of them has a planet or two around it, at least. All those little solar systems will have all sorts of random alignments, which means some of them will coincidentally be lined up *just so*, and as those planets make their lazy ellipses around their parent star, it will cross through our line of sight to that star.

Since planets are made of rock, ice, gas, and other various light-blocking materials, a little bit of starlight will be absorbed by that planet instead of making its way over to our telescopes. Which means the star will appear a little bit dimmer than usual. And when the planet moves out of the way, the star will return to Normal Brightness Mode.

A transit event. A sign of a planet outside the solar system. An extrasolar planet. An exoplanet.

There are other techniques the plucky astronomer will employ to detect foreign planets too, and the studious astronomer will use more than one method to cross-check results before making a big announcement. For example, as the planets loop around their stars, that star will wiggle to and fro from the gravitational pull of that planet. It's not a big effect, but if the orbiting planet is big and bulky enough (bonus points awarded for being extra close), we can measure it, either by literally watching the star wiggle or by looking at the shifting of light from redder to bluer as the star moves back and forth.

Heck, even Jupiter is able to make the sun move a distance greater than its own radius, so this isn't crazy.

Yet another approach relies on the subtle bending of spacetime from the stars and planets themselves. If light from yet another, even more distant, star passes through the system we're staring at (which is bound to happen every once in a while, because there are way too many stars out there), then that light will get distorted and magnified. This is all laid down quite not clearly at all in the mathematics of Einstein's general theory of relativity, and it's called microlensing.

Oh, and in some rare cases we can *literally take a picture of an exoplanet* by exceedingly carefully blocking out the light from the parent star. That method takes some serious astronomical guts, but it's possible.

All these techniques rely on random chance: the solar systems we're studying have to have the right alignments at the right times for us to capture a hint of another world. Still, we've managed to do this several thousand times over already, and will have heaps more as the decades roll on.

And what we've found would blow the minds of even the most creative science fiction authors. Seriously, we think we have decent imaginations, but the natural universe puts us all to shame.

Planets bigger than Jupiter. Planets smaller than Mercury. Planets bigger than Jupiter orbiting closer to their star than Mercury orbits the sun (we call these "hot Jupiters" because why not). Purple planets. Planets with rings. Planets with moons. Planets with moons and rings. Planets around red dwarf stars. Planets around white dwarf stars (this one was the first ever detected, way back in 1992, by the way).

Stars with a single planet. Stars with a dozen. Stars with multiple Earth-sized worlds orbiting in the Habitable Zone, so close that from one world you could easily see the surfaces of the others with just your naked eye (assuming you can survive on that world, but that's a mere technicality at this point).

Even our nearest next-door neighbor, a little red dwarf by the name of Proxima Centauri, hosts a planet.[17]

There are worlds out there, countless. And I'm not even mentioning the *rogue* planets; the planets with no stars to call a home, that just wander about the galaxy like . . . well, rogues, I guess.

How many planets in the Milky Way galaxy? Estimates vary in the range from "wow that's a lot" (a hundred billion) to "no way can that be true, but I guess we have to believe it" (a trillion).

We've just started taking a galactic census, but early estimates place the number of Earth-sized planets orbiting sunlike stars in their Habitable

Zones (in other words, the best places to look for life just like what we're familiar with) at around five. Billion. Five billion.

I don't know if we're alone, but Earth certainly has a lot of sisters.[18]

Now, the looking for life bit that you've been waiting for: when those planets cross the face of their stars, the light from that star trickles through the atmosphere of the intervening planet (if it has one), and light that passes through a gas (like, for example, the gas of an atmosphere) will change, very subtly. Different elements will absorb or emit specific frequencies of light, and by looking at how the light changes during the transit event, we can run the math backward and figure out what the chemical cocktail of that atmosphere is.

For example, we could look for . . . hmmm . . . how about . . . oxygen? Yeah, oxygen. O2 is highly volatile—it just doesn't like to stick around for long in an atmosphere. So how did the Earth end up with so much of the good breathable stuff in our air? Through photosynthesis. As in, life.

It's not a 100 percent surefire guaranteed *biosignature* (there's a word for ya), because of course nature can outwit us and come up with some fancy, convoluted process for generating a lot of oxygen on its own, but it's a major clue. A major sign of life on another world.

I'll repeat, to date we haven't found any signs of life, using this technique or any other, but we're trying. Hard. Someday soon somebody's bound to show up, as evidenced by a tiny little bump on some line chart in a random observatory.

With bigger telescopes we can go one up, looking for, say, methane (remember: fart gas), or CO2, or even industrial pollutants.

That's why we don't have to worry about broadcasting our existence to the wider galactic community. If they have telescopes and two brain-cell-equivalents to rub together, then they've already figured out this trick with the transit and the atmospheres. And Earth has had excess O2 for well over three billion years by now.

Like I said, if anybody's out there, *they already know we're here.*

★

I suppose you could try to leave the solar system and visit aliens yourself, which I've warned you repeatedly against, because of all the dangerous dangers involved in the journey, but if you're really dead set on it (literally) then who am I to get in your way?

No, the only thing getting in your way will be the ridiculous amount of space between you and anything evenly remotely interesting.

Unless you find the vacuum of space itself fascinating, in which case you'll be riveted for eons upon eons to come.

Aliens might or might not be out there (I think we've covered this in enough gruesome detail), but if they are, they are far away, in the most extreme possible interpretation of the word "far."

Space doesn't mess around. Just when you start to think "man, there's a lot of space," space smiles slightly and just keep on going. Think of the light-year. One light-year doesn't *seem* like very far, because there's only one of them, but we deliberately chose that word so we wouldn't have to keep repeating -illions when describing distances to the nearest star. It takes years for our fastest spacecraft to reach the planets of our solar system. At the same speed—blisteringly fast at tens of thousands of miles per hour—it takes . . . tens of thousands of years just to crawl past the outermost edges of our own Oort Cloud, let alone reach the vicinity of another star.

Tens. Of thousands. Of years.

Humans haven't even had *writing* for as long as it will take our fastest craft to reach another star.

If we ever find evidence for intelligent civilizations, communication itself will take hundreds, if not thousands, of years for a single back-and-forth handshake and courteous but noncommittal morning greetings. Visiting for supper is right out.

So why don't we just, you know, go faster?

I'll go ahead an answer my own question. Because energy, that's why. It takes raw, pure, unbridled energy with a capital E to accelerate something faster.

Consider, for example, one of the most plausible schemes developed to get a spacecraft going: the so-called Breakthrough Starshot initiative.[19] This plan calls for a we're-totally-not-mad-scientists-we-promise combination of a giant laser, reflective sails, and tiny robotic spacecraft. In the current reference design floating around such circles, the goal is to reach 10 percent of the speed of light. OK, that sounds nice. The spacecraft itself can't carry enough fuel (either gas or nuclear or anything else) to get up to those speeds by its lonesome, because more fuel means more weight, which means you need more fuel to push the thing, which means more weight, which means more fuel, and so on. Sigh.

So instead, let's keep all the fuel in one location, say on the ground, and transport the energy to the spacecraft to get it moving. With lasers. Because why not.

Really, because lasers are pretty decent at getting energy from Point Earth to Point Spaceship, so good in fact that they tend to melt whatever they're lasering at. Whoops. OK, so to counteract this and actually turn all that laser energy into movement, we'll attach to this spacecraft a solar sail—a highly reflective material (and by "highly" here I mean more reflective than any known substance, because if even a tiny fraction of the incoming energy gets turned into heat, it's game over) to bounce the laser light off of.

Light has momentum, and can push things. If this seems weird, then just don't worry about it.

So we shoot the laser at the solar sail, accelerating the spacecraft, sending it on its merry interstellar way.

To do this we need to have the largest, most powerful laser ever built—an array pumping out a total of around 100 megawatts (MW). That's like parking all the nuclear power plants in the United States right next to each other and hooking them up to the laser.

And the laser has to run for ten minutes, continuously, to accelerate the spacecraft to one-tenth light speed. That's about a billion billion times longer than our current most powerful lasers can operate.

Oh, and the spacecraft itself? With this much energy blasting it, in order to get to the intended speed it has to be lightweight. As in, about a gram.

That's a paperclip.

And it would reach Proxima Centauri in four decades.

None of this is strictly impossible. It's not like we're breaking any laws of physics here in our attempt to design a viable interstellar spacecraft. It's just . . . hard, man. Exhaustingly hard. Painfully hard. Annoyingly hard. So hard that sometimes it makes you want to stare up at the deepness of the heavens above and then go to bed.

I know, I know, this entire book is about interplanetary (easy mode), interstellar (medium mode), and intergalactic (hard mode) adventure and exploration. Or more accurately, a comprehensive and potentially exhaustive list of all the nasties you're likely to encounter. But to encounter them, and especially any potential life forms (whether simple and to-the-point bacteria or complex civilizations with lots of to-the-points aimed in your direction), don't ever forget that it's going to take a lot of raw energy to get there.

I don't know how you're going to design your spacecraft, and I don't know how long it's going to take you to get anywhere worth being there. Those are all engineering questions, which is quite importantly and absolutely Not My Department.

Thankfully, there is one tiny little saving grace, and that's a little quirk of the universe identified by one person you've heard of (rhymes with "Schmalbert Schmeinstein") and two people you haven't (Henri Poincaré and Hermann Minkowski), and it's wrapped up in a bigger picture of the universe known as special relativity.[20]

IF (capitalized for emphasis and not because my shift key got stuck) you're able to accelerate yourself to a good fraction of the speed of light (we're talking 99 percent and above here, for the record) then your travel times to distant and even more distant locales throughout the universe will be surprisingly short.

That's because the faster you move in space, the slower you move in time. Put another way: moving clocks run slow. Put yet another way: not everybody always agrees on the rate of flow of time. Put even still another way: to the outside universe it may take you tens of thousands of years to reach your destination, but to you it could be just a few weeks.

This strategy requires an enormous amount of fuel (uh, possibly more energy than you could ever hope to acquire from the entire lifetime of the sun), but again, that's not my problem. And please, for the love of stardust, make sure you don't ram into anything on your way out—even a tiny micro-meteoroid, when encountered at nearly the speed of light, has a *surprising* amount of kinetic energy, which you will have the full pleasure of getting acquainted with, for the remaining millisecond of your life.

Oh, and due to your frightening speed, and low-energy radiation just hanging around the cosmos (like the all-pervasive cosmic microwave background, which provides the vast majority of all radiation in the universe, but is usually and safely down in the low-energy bands of the electromagnetic spectrum) get boosted to high energies. X-rays. Gamma rays. Blasting you in the face. All because you're moving fast. Have fun with *that*.

Anyway, time doesn't feel weird when you're traveling fast—you just keep on being your normal self. But on the other side of the spacetime coin, distances appear shorter, meaning your destination appears fantastically nearby when you're booking along at these incredible velocities. Don't worry about the hows or whys; that's just how relativity works, and if it makes our adventures more palatable, then you bet your exhaust we're going to use it.

The upshot is that the incomprehensibly vast distances in our universe don't always have to be that way, if you're clever enough to go stupidly fast. So there's a chance that if you were to identify the existence of an intelligent life-form, you might be able to drop in for a visit, after all.

But during your near-light-speed travels, time still flows at its normal sluggish pace for the rest of the universe, meaning your newly discovered friends may be long extinct by the time you arrive at their planet.

And when you come back home to Earth, everyone you ever knew and loved will be gone, long gone.

That may or may not be a bonus perk, depending on your personal situation.

Wormholes

Mother nature loves you
Gives you warmth, light, a home
But don't test her.
She doesn't like a cheater.
> —Rhyme of the Ancient Astronomer

N o, I'm not going to talk about wormholes.

★

Fine, I'll talk about wormholes. But you're not going to like it, so this is your last and final warning to just skip this chapter altogether and read the final closing comments that come after this, which are full of hope and longing for discovery and adventure. It's all sweet and treacly and right up your alley, you starry-eyed wannabe wanderer of the cosmos you. Go on, you won't regret it.

Still here? Alright then, don't say I didn't warn you.

Wormholes don't exist.

There, I said it right up front, so we don't have to waste anybody's time. OK, OK, I'll be more pedantic so we can actually talk about it: wormholes have not yet been proven to *not* exist, but nature has been pretty clear to us so far, and every time we come up with a clever scheme to concoct one, in comes some Rule of Physics to smash our hopes against the rock of reality.

If in the extremely rare chance that you actually find yourself staring at a bare-naked wormhole entrance, just calmly back away without making any sudden gestures or baring your teeth. Wormholes take that as a sign of aggression. If you're lucky, you'll survive the encounter no worse for wear. If you're unlucky, at best you'll be atomized in less time than it takes for you to realize that you're in a wormhole. At worst, your component parts will be distributed throughout the known universe.

I have nothing against wormholes personally. It's not like they insulted my mother a long time ago or anything. And they're perfectly legitimate and sane solutions to the equations of general relativity. It's just . . . well, you'll see.

And I can totally empathize with all the travelers (which is *all* the travelers) who want wormholes to exist, really badly. Who wouldn't? Wormholes are shortcuts. Wormholes are cheat codes. Wormholes let you get from A to B without all the tediousness of actually having to travel from A to B.[1]

I know we talked about normal space travel, and how if you go really, really fast you can reach far-off destinations in a surprisingly short amount of time. But that trick comes at a tremendous cost: energy. Good luck generating and gathering and storing and harnessing all those ridiculous amounts of raw power needed to accelerate the likes of you (and all your stuff) to the nearest star, let alone somewhere worthwhile.

But what if I told you that you could reach distant, fanciful locations—the places of your very dreams—without a) using all the energy available in something like the sun, and b) without ever having to travel faster than the speed of light (which is a very hard no-no when it comes to the universe)?

You'd take the shortcut, every single time.

You'd take the wormhole.

Wormholes are a story of gravity, and gravity is a story of general relativity, so let's talk about the general relativity part first. Like I said way, way back in earlier chapters (Those were the days, we were so young then, weren't we?), GR (Gastroesophageal reflux? No. General relativity.) tells us about the relationship between stuff—planets, people, photons—and the bending and warping of spacetime. The dancers do their dance on the floor, and if there's a pit or curve in the floor, their dance will change, but the weight of the dancers causes those same pits and curves.

GR isn't just words: it's a bunch of math equations that describe this relationship. In fact, they're a set of incredibly difficult math equations, which is part of what makes gravity so "fun," but that's why we have professionals. Usually we're given some scenario, like a homework problem: a neutron star is orbiting a spinning black hole, how long until they collide? Given the setup, we can crunch through the GR math to figure out how everybody will act.

But it can also work in reverse. There are two sides to every equation, after all. Instead of taking the stuff and figuring out how spacetime will curve, we can imagine curving spacetime, and asking what kind of stuff it would take to make that happen.

And so: wormholes. They are an answer to a very particular question: can we bend spacetime *so much* that we punch a tunnel from one part of space to another?

In our universe with its three spatial dimensions (if you want a discussion on string theory and extra dimensions I suggest you petition my publisher to accept my manuscript "How to Die in Quantum Space"), the entrance to a wormhole looks like a giant ball floating out in the middle of nowhere. It looks a lot like a black hole event horizon (and in some cases it is, but we'll get to that in a little bit), but if the wormhole is working then that entrance looks like a magical crystal ball—when you stare at it, you see light pouring out from some distant corner of the universe. To enter the wormhole you . . . enter the wormhole. It's really that straightforward. Just walk in. There's no fanfare or exciting theme music. You just enter, travel down the tunnel (in the jargon favored by physics types the tunnel of a wormhole is called a "throat"), travel its length, and come out the other side, easy-peasy.

There's technically no restriction on the length of the wormhole. It could, for some unfathomable reason, actually be longer than taking the simple-but-exhausting straight-line path. But I'll assume for the sake of this discussion that we're only interested in the ones that make interstellar travel *less* boring.

But, gee, can it really happen? Just because we can imagine it doesn't mean that GR allows it.

GR allows it.

The equations of general relativity allow you to build a wormhole—a tunnel from one part of spacetime to another—if you can find a particular arrangement of matter to bend spacetime in just the right way.

But you know me by now—it ain't that easy.

What follows is a few various and sundry attempts at building a wormhole, starting with the simplest and progressively getting more elaborate. At every step, I'll show you how that particular attempt fails in an obvious or nonobvious way, and we'll move on from there. At the end of this, we'll have a relatively exhaustive, but ultimately unsatisfying, discussion of wormholes. Like I said before, they won't be *completely* ruled out, but the strains you'll have to put on our conceptions of reality will be so torturous that you'll probably wish they were never even thought of in the first place.

✪

The journey to the wormhole begins with a simple question: what if we took a black hole *and made it even more awesome*?

With words, it's easy enough to describe a black hole, which I did earlier to our shared delight and horror. In short, a black hole is a puncture in spacetime itself: a point of infinite density, where all the matter that has ever flowed in got crammed into an infinitely tiny point. We know that's not a completely accurate description—singularities aren't a thing, but rather a signpost that Einstein isn't going to guide us any further—but for now let's roll with it. These singularities are enveloped by an event horizon, which is not so much a *thing* as an invisible line in the sand. Once you cross the event horizon, you just got yourself a one-way ticket to Singularity Town.

Black holes are suspiciously easy to describe. With just three easy numbers (mass, charge, and spin) you know everything there is to know about a black hole. They don't make for great conversationalists.

And that's it. Three numbers to describe it. Singularity at the center. Once something falls in, it can't come out.

What we call a black hole is really a set of mathematical equations that describe the above sentences, and these equations are solutions to Einstein's general relativity. Karl Schwarzschild, remember him? Fun guy, flair for the dramatic.

But the equations that Karl and company solved to arrive at a black hole can be flipped around: you can toss some minus signs into the math while keeping everything consistent. GR perfectly and without complaint allows you to construct the Good Twin to the black hole: a singularity, surrounded by an event horizon, but where nothing can ever fall *in*. It's entirely impossible to cross the event horizon. Quite the opposite: stuff is constantly flowing/flying/blasting (choose your preferred word here) *out* of one of these objects.

Meet the white hole.

In fact, if you take the math of GR seriously (which you should), then it turns out that every single black hole is paired with a white hole, with their singularities kissing:

Black hole event horizon—singularity—white hole event horizon

What's even stranger, once you sort out where all the minus signs needed to go to make the math balance, and follow them through to the end result, you end up with a situation known by several names. Here are some: 1) Schwarzschild wormhole, 2) eternal black hole, 3) Einstein-Rosen bridge.

In this first, and simplest, construction of a wormhole, the white hole always sits in the past, and the black hole is always in the future. The white hole has its own event horizon, but you don't really care about that one because a) you can't get into it anyway, and b) it's in your past. What's in your future is the black hole event horizon, which thanks to its joining with the white hole has made a copy of itself, with one horizon in your neck of the woods and the other horizon somewhere far out in space.

And there's a bridge. A region of spacetime connecting two different places in the universe. The act of joining up a white hole with a black hole makes an entirely new structure, where two different observers (let's go with Alice and Bob, because they are our steadfast go-tos when it comes to black hole spelunking), both staring at their versions of the black hole event horizon, and can journey into that in-between place.

So Alice and Bob hop in.

And immediately die.

Technically, the game of the wormhole has been achieved: a tunnel connecting two regions of the cosmos. But also technically, Alice and Bob passed through an event horizon of a black hole in order to make this journey.

And what happens when you pass through the event horizon of a black hole? Singularity. It's unavoidable. It's literally mathematically impossible to escape a black hole, *because that's the definition of a black hole*. You cross that one-way barrier? You're doomed. No ifs, ands, or, especially, buts.

Sure, Alice and Bob get to meet, chat, share some tea, and exchange a brief but assuredly fascinating discussion on the nature of their respective regions of the universe. They can also see any light from both portions of the universe that's fallen in after them. And then they get crushed into oblivion at the singularity, because that's what black holes do, whether they're connected to a white hole or not. No, you don't get spat out the white hole. I'll repeat: you crossed the event horizon of a black hole. You're done. Goodbye.

The only way to escape such a wormhole is if you happened to be inside the white hole when it formed. White holes have the opposite rule as black holes: nothing is allowed to stay in, so you get expelled, and it's your choice whether to travel to Alice's or Bob's place in the universe.

But don't worry, my little traveler. These kinds of simple wormholes most certainly don't exist in nature.

For one, white holes. Sigh. We know how to make a black hole, and nature does it just fine on the regular: scrunch down a bunch of stuff below a certain threshold, then gravity takes over and does its gravitational thing, which is to keep pulling. And there you go: a singular point of infinite density surrounded by an event horizon.

But a white hole? It still needs a singularity—a point of infinite density, but the matter needs to flow from it rather than toward it. Yes, it's *technically* allowed by the equations of GR . . . just like all the little water droplets in a lake can, suddenly and spontaneously and without anybody asking, decide to flow up the cliff, making a reverse-waterfall (a "waterclimb?").

Seriously—there's nothing in either Newton's or Einstein's equations of gravity that prevent this. But Newton and Einstein aren't the only people telling nature what to do. There's more than the force of gravity in the world; for example, there's thermodynamics, and thermodynamics have some hard and fast rules about the way things ought to be.

There's nothing stopping a single random water molecule from jumping up the cliff face. But two? A hundred? A trillion? The entire lake? The math of thermodynamics (specifically, entropy) tells us that the *chances* of that happening are so extreme that it makes the astronomical numbers we've been using in our exploration of the cosmos seem downright comical in their smallness. There are small chances, and then there are thermodynamically small chances.

The upshot is that in your entire lifetime, or the lifetime of the universe, or a million lifetimes of a million universes repeated millions of times over, you're just not going to see a waterfall go the wrong way.

And you won't see a white hole, either. It has attractive gravity, but all the material in it has to decide to flow away from it, against that pull.

Good. Luck.

Anyway, the recipe nature uses to make black holes (collapsing stars) simply doesn't make a corresponding white hole—the presence of all that star stuff simply snuffs out the formation of the white hole before it even has a chance to wreak havoc with our sensibilities. The mirror image to a black hole appears to be just another fluke of the mathematics; a ghost with no corporeal form in the equations.

Besides that, if a white hole were to *somehow* exist, then it would be destroyed by nothing more than a single particle. That particle would fall toward the white hole, because gravity, but could only ever get closer and closer to the event horizon. This increases the energy of the particle to no end (because it gets closer and closer but never reached the horizon). And what happens to particles when they have near-infinite energy? You don't want to find out.

So that's the wish-you-were-here white hole. But even if you got a white hole, violating physics all willy-nilly, you *still* couldn't make a wormhole. The problem is the singularity—the same singularity that spells your doom as soon as you enter the wormhole. Singularities are places of—to put it mildly—extreme gravity. Do you think the tunnel of a wormhole can stabilize itself against the horrific gravitational stresses and forces acting on it near the singularity?

The answer is no, you don't think so. As soon as this simple kind of wormhole opens, the tunnel immediately collapses in on itself and shuts, like a rubber band stretching itself so thin that it snaps. How quickly? Faster than even a beam of light could travel from one end to the other.[2]

No white holes. No stability. No wormholes. Is this the end of the interstellar road?

✪

To be fair, we started with the simplest, most bare-bones kind of wormhole possible. The black hole envisioned by Schwarzschild doesn't have any electric charge and isn't spinning. That's fair, let's pick on something our own size.

Both electrically charged black holes and their rotating cousins share one curious property: they automatically, without any hesitation or flinching, create a wormhole. What's better, these wormholes work way better than their more straightforward relatives described earlier. There's still a white hole anchored to the black hole by their singularities, but in this case you can enter the event horizon of the black hole, travel near the singularity *without getting swallowed by it*, and find yourself expelled out the white hole in another universe.

I know, I know. I just got done repeating that once you cross a black hole event horizon, there's no going back. But that was for *simple* black holes. Now we're dealing with *more sophisticated* ones. Different situation. Different math. Different results.

This works because in both the case of the charged-up black hole and the spun-up black hole, there are some extra gravitational effects that we have to keep track of in our spreadsheet of doom.

Remember that the singularity is a point of infinite density. It's where all the things go when they fall into a black hole. All the mass. All the charge. All the spin. That's where it goes, compressed into an infinitely tiny point.

You take a bunch of mass and you cram it down into a point, now you have a region of extreme gravitational force.

You take a bunch of electric charge and cram it down into a point, now you have a region of extreme electric force.

You take a bunch of spin and cram it down into a point, now you have a region of extreme centrifugal force.

You see my point?

When you fall into an electrically charged black hole, at first nothing special happens (except for the fact that you just fell into a black hole). Gravity continues to pull you toward the singularity faster than the speed of light, and it just keeps on getting stronger. But as you approach, the insane electric field starts to kick in. That electric field has energy, right? Right. And the bending of spacetime responds to matter and energy, right? Right. And the bending of spacetime tells other things how to move, right? Right.

The energy in the electric field has its own gravitational effect, and that effect is repulsive (because the math said so).[3] So as you approach the singularity and prepare for your ultimate demise, you find yourself slowing down, stopping, and reversing, the antigravity of the strong electric field *propelling* you faster than the speed of light.

You are pushed and pushed until you pop out the other end, beyond the event horizon of the white hole, in some random part of the universe. You don't ever get to go back to the place you started (unless you take the tedious route), because black holes are places where space flows inward, and white holes are places where space flows outward. You can't have space flowing in both directions at once, so the output-end *must* be different than the input-end. You pass through the wormhole, never touching the dreaded singularity, and go on your merry way, no big deal.

The rotating black hole is the same, but with the intense spin causing a centrifugal force that has the same effect as the electric field: a repulsive antigravity deep in the heart of the black hole. (There's also a part here about the singularity getting stretched to form a ring, and a region around that ring that allows time travel, but let's not worry about that.) It's possible for a curious traveler to pass through that ring, experience the fun and thrill of antigravity, and get squirted out the other, whiter, end of the wormhole.[4]

Easy-peasy, right?

You won't be surprised when I tell you—as I will continue to tell you at every single opportunity, so don't even bother getting your hopes up—that this doesn't work.

First, charged black holes. If a black hole were to form with some electric charge, it would be especially attractive to the opposite charge, sucking down anything it can, as fast as it can, to rebalance its internal attractiveness. And since the universe is overall electrically neutral, it's highly unlikely that a black hole will be born charged one way or another.

Still, not technically impossible. Just unlikely. But that's not enough to kill a black hole.

Besides, rotating black holes are the *norm*. Just about everything in the universe is rotating, and if it isn't already it's going to soon. Black holes form from stars, and stars are spinning things, so black holes are automatically spinning things, too.

The default state of a black hole in our universe seems to be: massive, uncharged, and spinning like crazy.

But as sensible as these black holes seem (for certain curious definitions of the word "sensible"), things go haywire on the insides when we examine them in more detail.

Remember that part, both in charged black holes (which we don't really care about, but we'll include for the sake of manic completeness) and rotating black holes, where you were rushing inward toward the singularity, slowed down, stopped, and reversed? That was just your experience, but what about the experience of everything else that fell into the black hole? All the matter, all the radiation accumulated over the billions of years of the black hole's existence?

From your perspective falling in, all the stuff that fell in previously is sitting there, waiting for you to get close enough (remember that gravitational time dilation is still a thing). When you get near, the entire past history of the black hole rams up into your face, hitting you with infinitely blueshifted, infinitely high-energy radiation.

And in the weirdness of white holes, on your way out you get a second blast, this time from all the matter and radiation *in your future*.

This, as you might imagine, causes some problems. If even a single particle falls into this kind of wormhole, the back-and-forth sloshing at this transition point reaches a fever pitch, wreaking gravitational havoc and ripping the wormhole apart at the spacetime seams.

It's, you guessed it, unstable.

But wait! I very clearly said that rotating black holes are a common feature of our universe. And they certainly exist, so isn't what I'm saying wrong?

Yes, what I just said is wrong. Obviously rotating black holes don't automatically tear themselves apart in a furious blast of radiation and matter.

Equally obviously, we have no idea what really goes on inside a rotating black hole. The math of GR is good enough to describe what happens outside that kind of black hole, leading to all the frame dragging and gut-punching jets that we are cautious enough to avoid. But the centers? That's a different story—a story that hasn't been written yet.

It's the same as the singularity itself, a word we toss about with careless abandon as if we actually knew something. The truth is, the centers of all black holes, charged or neutral, spinning or sedate, are mysteries of modern science. The whole wormhole thing seems doubtful, because of that ferocious instability, though. What likely happens is something equally violent but less useful for traversing vast interstellar distances.

You probably just get smashed.

✪

Again, we see the math of GR leading us astray, lulling us into a false sense of hope that this crazy wormhole thing might actually pay off, only to be confronted with Some Other Physics that enjoys ruining our party. Just because Einstein said so, doesn't make it so. Jerk.

But is there anything we can concoct to make this work? Scientists are smart folks, right? Surely *somebody* has come up with a workaround after all these decades.

Sure, why not? If we're in the land of make-believe physics, we can make anything happen. We see now that the challenge isn't necessarily how to make a wormhole, it's how to make a *stable* wormhole, one that we can cross from end to end, reaching our dreaming destination, without getting turned into a human puzzle piece.

Oh, and there's that whole part about needing to avoid the singularity. Right. In the jargon that will come in handy when you're trying to smooth talk your way through a reception at the next "Black Holes and You: A Galactic Symposium" meeting, we're looking for "traversable" wormholes.

It turns out that there's one special trick that can do both jobs simultaneously: holding the wormhole entrance away from the event horizon *and* maintaining the stability of the tunnel itself as you rabbit your way through it. Amazing! Of course, this two-for-one miracle product comes at a hefty price (hint: it may not exist).

All you need to turn all your wormhole fantasies into reality is a nice healthy lump of negative mass.[5]

Negative what?

Negative mass.

It's mass, but negative. Like, something that weighs negative five pounds.

It sounds preposterous at first glance, but let's not blow it off completely. After all, it can make wormholes an actually viable thing, so let's explore it a little bit.

Negative mass has two important properties that make it useful for building a wormhole. One, its very negativity alters the structure of spacetime around a black hole, pushing the entrance to the wormhole outside the event horizon, meaning that you can jump into the tunnel without ever crossing that treacherous one-way barrier of doom—and since you never pass through the event horizon, you never have to worry about inescapably hitting the singularity. Nice!

Next, negative mass has an enormous amount of pressure, enough to force the tunnel open like the supports of a mineshaft. The amount of negative mass you have determines how wide your wormhole can be, and how much normal matter you can send down the throat before risking catastrophe.

In case you're wondering, and I know you are, to make this work, your material with negative mass has to be in the shape of a shell, to make the entrance to the wormhole in three dimensions.

It's a shame that negative mass doesn't exist.

Here we are yet again. The equations of general relativity don't really care one way or another if your mass is positive or negative. They just plug along doing their mathematical thing. And Albert is quite clear: give him some negative mass, and he'll give you a wormhole. But garbage in, garbage out. It's up to us to decide if what we're putting into the equations make any sense whatsoever.

And does negative mass make any sense?

I'm not being rhetorical—the question has come up for about as often as we've had a concept of "mass" (which by now is a long time). I mean, come on, there are both positive and negative charges, so why not positive and negative masses? We haven't ever seen any around, either here on Earth or in any of our interstellar travels or observations. But that doesn't mean it *can't* exist.

The definition of "mass" itself can get a little tricky depending on which physicist you're talking to and what their current mood is, but negative mass would be pretty funky. If you give a negative-mass soccer ball a good swift kick, it will go flying in the opposite direction. If you throw a negative-mass rock at the ground, it will go soaring up into the sky.

If you have two bits of negative mass, they will repel each other. If you put a positive-mass ball next to a negative-mass ball, the positive-mass ball will pull on its negative neighbor, while the negative ball will push on the positive on. The simple act of placing them next to each other without any constraint sets off their motion, which will continue accelerating to infinite speed.

If you put a positive mass on the opposite side of a wheel as a negative mass, the wheel will spontaneously start spinning and continue spinning forever, allowing you to take extra energy from such a device perpetually.[6]

This . . . seems a bit worrying. Don't we have rules for this universe? Like conservation of momentum and energy? The reality of negative mass seems so nonsensical that we're tempted to just outright outlaw the whole thing.

But we have to be careful here. There's no law of physics itself that rules out negative mass—it's just that negative mass violates what we currently understand to be *other* laws of physics. So which is it: can negative mass exist, and we need to update our understanding of momentum and energy to accommodate it? Or were we not kidding at all about those ultra-important and ultratested conservation laws, and we just need to toss negative mass in the cerebral trash bin, where it belongs?

If you want to believe in negative mass, I'm not going to stop you. But conservation of momentum is one of the most well-tested, well-understood (seriously, it's been put through the ringer a time or two) concepts in modern physics. You rely on conservation of momentum every single moment of every single day.

Negative mass is like negative food. It may seem exotic and thrilling at first, but it won't satisfy you in the end.

★

But the story isn't over yet. There's negative matter, which seems to be met with a resounding and confident, "No," from the community of people who think about such things. And then there's negative *energy*, which as we're about to find out, garners a much milder, "Uh, hmmm," in response.

General relativity cares about both matter and energy—if you have matter, you can bend spacetime to your will. And if you have energy, you have equal power. After all, the special kind of relativity taught us that $E = mc^2$. Energy *is* mass, with the speed of light (c) tossed in to make all the unit conversions work. Mass and energy are simply two sides of the same relativistic coin.

So with negative mass out of the picture (let's hope), here are the questions that confront those of us who want to build a wormhole: can we make negative energy happen, and if so, can we use that to build a wormhole? If it sounds like we're getting a little desperate here, it's because we are.

Once you open up the negative energy can of worms, there's no going back. And the very first worm to pop out is something called the Casimir effect, where two metal plates, when held very closely together, will feel an attraction to each other not through any force of nature, but because of the negative energy created between them.

It sounds like new age mumbo jumbo, but it's not nonsense; it's in that category of weird-but-true aspects of our universe. Which, now that I think about it, makes up most of this book.

To talk about the Casimir effect, we need to talk about those quantum fields again. I should've asked you to bookmark a few pages in the chapters about the vacuum of space and cosmic strings, because it's kind of important now. The short version, in case you slept through and/or skipped over those discussions, is this: the entire universe is filled with fields, and sometimes bits of these fields pinch off to become normal everyday particles.

Those fields have energy: they are vibrating, humming, buzzing. Even empty space is filled with these energetic fields. Just like the air around you is filled with sounds; as I type this I can hear rustling in the kitchen, the neighbor's annoying lawnmower, traffic down the street. The air is alive with vibrations of all kinds.

Empty space is constantly humming with vibrations of these quantum fields. They don't really *do* anything in their ground state. You have to add extra energy to them to get them to become particles and participate in physics. But they exist. And technically, as we explored to our dismay and confusion before, there are an infinite number of vibrations in the quantum fields happening all around you, all the time, nonstop. We don't notice or care about that infinite sea of energy that we're sitting on top of, because physics is a game of differences, and you can have our universe sitting on top of an infinitely high-energy mountain because that's just the way things are.

Key takeaway: infinite numbers of vibrations of every single possible wavelength happening in the vacuum of spacetime. Deal with it.

Let's say I pick up a tuba (why not?) and start a-blowin'. I can only make certain kinds of vibrations—the shape of the tuba only allows specific kinds of sound waves to form. We call these specific wavelengths of sound waves "notes." That's the whole point of building and playing a tuba.

Outside the tuba: all kinds of sound waves of any sort of wavelength. Inside the tuba: only specific notes can be played.

Empty space: all kinds of waves of any sort of wavelength in the quantum fields. But if I take two metal plates and hold them close to each other? Good job, you're paying attention: only certain kinds of waves will be allowed. The plates can only play certain specific "notes" in the fundamental quantum fields of our universe.

Well, so what? So what is that outside the plates you have many more waves than inside the plates (if you really want to fry your brain, read the rest of this parenthetical: technically, there are still an infinite number of different wavelengths that can exist inside the plates, but that infinity is slightly less than the infinity of wavelengths outside the plates, and we can play some very tedious mathematical games to extract the difference between those two infinites).

The end result is that with so many more waves outside the plates than inside, the space between the plates has literal negative energy. That negative energy manifests as an attractive force, pulling the plates together.

This isn't a children's fantasy. Earth scientists have done countless experiments to test this. It's real all right, and the effect is named after the guy who first figured it out: Hendrik Casimir. Negative energy is real and is here to stay.[7]

OK, maybe we're on to something! If we can create tiny amounts of negative energy in the laboratory with our parallel metal plates, could we use this to warp the geometry of spacetime in just the right way to open a portal to another part of the universe?

So there you have it: your Instant Wormhole Stabilizer©. All you need to do it take two metal plates, put them close together . . . and . . . uhhh . . . put it near a black hole? I guess?

First, the negative energy created by the Casimir effect is so tiny that the word "minuscule" seems gargantuan, and the amount of negative energy needed to stabilize a wormhole is so huge that the word "gargantuan" seems minuscule. They're just on totally opposite ends of the scales. And every attempt you can think of to scale up the Casimir effect just ends up making the negative energy even weaker.

I'll give you the benefit of the doubt, and I'm being *really* generous here, and let's say that actually constructing a wormhole with some sort of Casimir effect device is just something for the engineers to figure out. The physics are sound, and that's all that matters.

But are the physics sound? Like, do we actually know what the heck we're talking about?

The big question is the nature of that negative energy. Yes, the energy between the metal plates is negative *relative to the outside world*. It's not negative in an absolute sense. It's like climbing to the top of Olympus Mons, digging a hole one foot deep, and claiming that you're now below the Martian equivalent of sea level.

What's more, the space between the plates isn't the only game in town in the Casimir setup. There are also, you know, the metal plates themselves.

Which have mass. Positive mass. Which means positive energy. When you spreadsheet the entire Casimir apparatus as a whole, you end up with a net positive energy.

All this means is that it's a little unclear if the negative energy in the Casimir experiment is the *right kind* of negative energy needed to open and stabilize a wormhole. It very well could be, but we don't have a consensus on it yet—as you can tell, there's a lot of arguing and scoffing back and forth in the scientific community about all these technical definitions.

✪

Who knew that a discussion starting on the possibilities of wormholes would end up veering off into arguing over the technical definition of "energy?" But that's what we get when we try to cheat Mother Nature at the whole thou-shalt-not-travel-faster-than-light thing. Sometimes nature can just be a real pain in the neck, and there's nothing we can do about it.

Give theorists enough time and they will naturally expand any given open question into a whole branch of physics. Case in point: negative mass and/or energy. To generalize the discussion, various scientists have proposed so-called "energy conditions" on our equations to help guide our thinking.[8] They don't come from any other fundamental stance or hard experiment—they're the mathematical equivalent of throwing your hands up in the air and saying, "I quit. Negative energy doesn't exist."

But it's such a slippery subject that most scientists avoid the whole mess altogether and just go about tackling less challenging problems, like how stars blow up when they die.

Altogether, it looks like Casimir isn't going to save wormholes, even though he tried really hard. In the decades since it was realized that you need negative mass and/or energy to stabilize a wormhole, scientists have concocted all sorts of situations to make one stick, like using cosmic strings. Remember those ghastly beasts? It might (with a strong and healthy emphasis on the word "might") be possible to thread a cosmic string through a wormhole, and let it loop back on itself through normal nonwormhole space. Closed cosmic strings vibrate like your hands after your fourth shot of espresso, and those wiggles might just have the right interaction with spacetime to make the local energy dip into the red, providing the necessary stability.

At least, until the cosmic string vibrates itself into oblivion through the emission of copious gravitational waves. And who knows what havoc those gravitational waves will play on the wormhole throat itself. And you need another cosmic string threading the wormhole and extending out either end to infinity to anchor the whole thing in place, and . . . you see where these kinds of discussions wind up.[9]

Another possibility is the fundamental vibrations in the quantum fields themselves, the same ones that give rise to the Casimir effect. Sometimes playfully called the "quantum foam," spacetime at that sub-sub-sub-sub-subatomic level is a frothing, seething mess. Surely, sometimes a randomly wormhole-looking thing pops into existence before simmering back in the nothingness from whence it came. Maybe (I know I'm reaching but work with me here) we could catch one of those teensy-weensy wormholes when they form and inflate them to not-tiny sizes? This solves the question of how to construct one in the first place without a handy nearby black hole, but the issue of maintaining the stability of the now-enlarged wormhole is definitely somebody else's problem.

Maybe we're getting gravity wrong altogether! I mean, we're taking Albert at his word, and he seems like a trustworthy guy who did all his homework, but who died and crowned him the king of gravity? Oh right, Newton did. Anyway, general relativity certainly isn't the be-all and end-all when it comes to gravity: that's the lesson that singularities teach us. Would an upgraded version of our understanding of gravity permit the existence of wormholes? It might—but so far all our attempts to outwit old Einstein have failed disastrously, so good luck with that.

The big catch with all this is that we simply don't fully understand how quantum fields interact with gravity. Earth scientists have been playing the game long enough that they've figured out how quantum fields behave in curved spaces . . . kind of. The math isn't easy, and our view of the physics is . . . let's call it murky. And that's in *barely* curved spaces. Extreme places like black holes or wormholes? You can speculate all you want, make all the approximations and assumptions you can so that the math doesn't blow up in your face, but the sheer exoticness of the situation means that there aren't any handy-dandy experiments laying around for you to test your wild ideas.

✪

There's one other thing that I need to mention about wormholes before I drop the topic altogether, and this thing that I'm about to mention is either the most awesome thing you've ever heard, or the most terrifying.:

Wormholes can act like time machines.

Sigh, here we go.

Let's assume you can build a wormhole. Stable, avoiding the singularity, traversable, blah blah blah.

Take one end of the wormhole, put it inside your spaceship.

Accelerate your spaceship to close to the speed of light. Cruise around for a while. See some sights. Enjoy yourself.

Come back to the other entrance.

Drop off your end of the wormhole.

Time machine.

How? Because when things move really fast in space, they move really slow in time. Your journey amongst the stars at hyperfast speeds may have only lasted a few days, but to the rest of the universe, stuck in the relativistic slow lane that we are, weeks or even years could've passed.

So now you've got one end of a wormhole that exists in the "future" of the other end, meaning that you could enter the "future" end, casually stroll through the length of the wormhole, and arrive at your own past. You traveled in time, without violating any physics.

What are we supposed to make of this? To some, this is the gravitational holy grail: if we can prove that wormholes are allowed in our universe, then this means—by some very clear-cut math—that time travel into the past is also allowed, which is kind of a burning question in physics nowadays (and also ever since there's been a physics). Time travel into the past *seems* to be forbidden—we certainly don't have any evidence of it ever occurring—but we don't know why. Maybe the ultimate answer is hidden behind the inscrutable math of the wormhole.

Or maybe this is the ultimate signpost that yeah, no, wormholes don't exist. Maybe there's a reason—a deep, fundamental truth to our universe—that time travel into the past appears to be impossible. Maybe it . . . simply is impossible, for reasons that we haven't quite fathomed yet. And if time travel is strictly verboten, and the existence of a wormhole allows you to build a time machine, then that must mean that wormholes are a concept DOA.

This has led some theorists to propose that nature will always (*always*) figure out a way to snuff out whatever wormhole contraption you concoct. For example, it could be that as soon as you build your wormhole and do your little race-around-the-galaxy thing, particles will start zipping back and forth through the throat of the wormhole, colliding and interacting with their past/future selves, driving up the energies until the wormhole blinks out of reality before you even . . . blink.

★

Look, I'm as exhausted as you are at this point.

Here's the point, as if I haven't driven it home enough already: we don't understand the physics that would govern a stable wormhole. Right now, they are purely theoretical constructs (aka mathematical fantasies). Sure, they're *allowed* by general relativity. But that doesn't mean that they *must* exist. For a while, black holes were in a similar boat: allowed by the equations but unknown if they really existed. It could be that wormholes are the same, although they have the disadvantage of not having any known natural process in the universe that can create them. Bummer.

I can sympathize with wormhole wannabes. It's such a tempting idea: shortcuts through spacetime, avoiding the dullness and drudgery of space travel. Faster than light! Connect up two points with a wormhole, and you can shoot through it faster than light will take going the long way around. Maybe even time travel, if you arrange the holes right. Amazing! A cheat code for the universe!

Some of you may have encountered a shifty-looking vendor or two trying to hawk a so-called Alcubierre drive, which they claim allows you to travel to far-off locales faster than the speed of light, but without breaking any precious rules of the universe, simply by manipulating the fabric of spacetime around you. Guess what? In order to make an Alcubierre drive go, you need to stuff your spaceship with negative mass. It's the exact same predicament as the wormhole—if they're possible, we have to completely rewrite our understanding of the universe.

It seems, as far as we can tell, that nature simply doesn't want us to make wormholes or warp drives or anything else like that, for inscrutable reasons known only to her and her alone. We're just going to have to stay in the dark. Literally, because we'll be spending the vast majority of our time in the cold night between the stars.

Even if you spotted a wormhole, I wouldn't jump in. It may do what you want it to do: jump you to another part of the universe without breaking a sweat. It may also explode. Or implode. Either way, it will most likely kill you instantly and horribly, spreading your parts all around this universe and probably the universe next door. I mean, seriously, these are extremely hypothetical objects that, even on the best of days, can hardly be described as *reliable*. Are you seriously going to risk life and limb for that chance at convenience?

Stop, I don't want to know your answer.

In a weird/beautiful sort of symmetry, I started our tale by talking about the cold hard vacuum of space itself, that the most dangerous part of space was all the space. But all that space held a secret: a frothing, vibrating background humming to the universe itself, thanks to all those elusive quantum fields. And here they are again, taking center stage in the discussion of whether it's ever possible to avoid all that blank emptiness between the stars.

I understand your desire to explore all these exotic destinations, despite all my protestations and carefully worded warnings. And I don't blame you for wanting to get to those destinations as soon as possible. Wormholes and warp drives seem too attractive, and always seem right at the cusp of being possible. But for the time being and time to be, it looks like that in this universe, if you want to get around, you have to do it the hard way.

EPILOGUE
A Final Warning

In the end, all that mattered
was that some matter
learned matters most important:
the curious mind never rests.
                —Rhyme of the Ancient Astronomer

There, that about does it.

That wasn't so bad, was it? Oh wait, it was bad. Very, very bad. Turns out there's a heck of a lot of stuff in this universe that can whack you without even batting an eye. They come in all shapes and sizes: tiny rocks that can punch a hole in your ship. Gigantic black holes the size of solar systems chowing down on anything they can reach. Microscopic particles that mess with your DNA. Blobs of charged gas hurtling through space.

There are a few common themes to these dangers. Gravity plays a big a role: any place where things can fall "down" usually heat up and start causing trouble. Magnetic fields had a surprise breakout performance: usually we hardly notice them, but make them strong enough, and they'll hurt you with a vengeance. While most of the universe is a dull empty lifeless wasteland, all the actually interesting parts are full of so much energy that our fragile human bodies and human-built ships just aren't up to the task of coping.

I think I covered the most important bits, as far as our limited understanding can take us. But understanding is always limited. Observatories on planets, slow-moving probes, and satellites can only tell us so much. Who knows what other dangers lay out there in the uncharted vastness of space? I'm sure you'll find out, or at least somebody else will. The question is: will they live to tell the rest of us about the wonders they have witnessed?

I hope you've been thoroughly warned, enough so that you'll seriously reconsider a career as "space explorer," because that quickly turns into "remembered as a space explorer." Take a deep breath. Doesn't that air feel good as it fills up your lungs? Stretch out your arms. Look at all the room you have! Go for a walk on a sunny afternoon. Isn't it comforting to have the constant reassuring pull of gravity beneath you?

Space isn't for everyone. No matter how clever our spaceship designs become, no matter how much we invest in new technologies, it will never be natural for us. We were born on planets, our parents were born on planets, our parents' parents' ever-so-distant parents were born on planets. Space exploration is a hard, nasty business. It's full of mind-numbing nothingness dotted with brief moments of ferocious intensity. It's not for the faint of heart or the weak of spirit.

Although . . . I have to let you in on a little secret. The universe is a dangerous place, that much I've made abundantly and painfully clear. But it's also beautiful. Sublime. Exotic. Wonderful. Thought-provoking. It's a canvas painted with the brilliant hues of matter and energy. Physics is the brush. For centuries the heavens have called to us. As we peel back the mysteries new ones emerge.

The universe is *begging* us to explore it. It's called to you; can't you hear it? There's only so much we can learn, bound to planets as we are. We need to go there. To shove our hands into new dirt, to let foreign light into our eyes. To learn, to understand, to *sense*.

I wrote these chapters to weed out the weak and unwilling. To scare some sense into them. For the remaining, the more foolish and daring and curious than usual, this book is a guide. It's really an excuse to talk about all the wonderful physics happening in the cosmos. The physics I described is known but, in most cases, not very well. There is so much to learn, and we need to study it as closely and intimately as possible.

Be wary of danger. Be cautious in new systems. Tread carefully in unfamiliar galaxies. Take your time; it's one of the few things you have on your side. There's a beautiful and wonderful universe out there, and it's yours. You may have to share it with others—people or otherwise—but it's plenty big for all of us, and getting bigger every day. We've haven't even scratched the surface of what reality is and what it contains. Every day is a new surprise. We'll never have complete understanding of time and space, but the journey is the fun part anyway.

Go, explore!

ENDNOTES

THE VACUUM

1 As usual, it all starts with Aristotle writing down everything he thought of, whether it made sense or not: Aristotle, *Physics*, Book IV.

2 I'm not joking, and neither was he. In the mayor's own words: Otto von Guericke, *Experimenta Nova (ut vocantur) Magdeburgica de Vacuo Spatio*, (Amsterdam: Johann Jansson, 1672).

3 And for our purposes we can gently smooth over the fact that Kepler was a nutcase, and I say that in the most endearing way possible: Johannes Kepler, *Harmonices Mundi*, (Liza, Austria: Johann Planck, 1619).

4 And nobody could go Over It like Newton, who had to invent an entirely new branch of mathematics (calculus) just to prove his point: Isaac Newton, *Philosophiae Naturalis Principia Mathematica*, (1729).

5 Back in the day, Earth scientists would get major acclaim for simply lecturing on the subject of, "Hey guys, check out what I tried last week": Thomas Young, "Bakerian Lecture: Experiments and Calculations Relative to Physical Optics," *Philosophical Transactions of the Royal Society*, 94, (1804).

6 Sadly lacking in any beard-grooming tips: James Clerk Maxwell, "A Dynamical Theory of the Electromagnetic Field," *Philosophical Transactions of the Royal Society of London*, 155, (1865).

7 And that nothingness was of a truly extraordinary kind. Take that, ether: Albert A. Michelson and Edward W. Morley, "On the Relative Motion of the Earth and the Luminiferous Ether," *American Journal of Science*, 34 (203): 333–345, (1887).

8 For some fun homework exercises to amuse you on your journey and/or help you get to sleep, try this one: Tom Lancaster and Stephen J. Blundell, *Quantum Field Theory for the Gifted Amateur* (Oxford, UK: Oxford University Press, 2014).

9 You can even use online calculators to map this out yourself, or to provide estimates for any systems you might encounter. Assuming you get a decent Internet connection: "Stellar Habitable Zone Calculator," (http://depts.washington.edu /naivpl/sites/default/files/hz.shtml), University of Washington, retrieved November 5, 2019.

10   Not many people get to define something in their own autobiography, but this dude managed it, and got the Karman line named after him: Theodore von Karman with Lee Edson, *The Wind and Beyond* (Boston: Little, Brown and Co., 1967).

ASTEROIDS AND COMETS

1   What is it made of? Smashed-up planets. Remnants from explosions long ago. Maybe even the ground-up spleens of some ancient alien species. Now you need a shower, don't you? *Accretion of Extraterrestrial Matter Throughout Earth's History*, Bernhard Puecker-Ehrenbrink and Birger Schmitz, eds., (Berlin: Springer Nature, 2019).

2   It's OK if you don't believe me. NASA doesn't care: "Micrometeoroid and Orbital Debris (MMOD) Protection," (https://web.archive.org/web/20101226102151 /http://www.nasa.gov/externalflash/ISSRG/pdfs/mmod.pdf), retrieved November 5, 2019.

3   Some astronomers even spend their careers trying to clean up all this dusty history. I gues it pays the bills: "Stochastic Histories of Refractory Interstellar Dust," *Lunar and Planetary Science Conference Proceedings* 18 (1988).

4   When even serious academic researchers are tossing around words like "cataclysmic,", you know it's bad news: R. Gomes, H. F. Levison, K. Tsiganis, and A. Morbidelli, "Origin of the Cataclysmic Late Heavy Bombardment Period of the Terrestrial Planets." *Nature* 435 (2005).

5   Yes, it was just a rock and not a visiting emissary of any alien species. Also yes, that's cool enough on its own: K. J. Meech, et al., "A Brief Visit from a Red and Extremely Elongated Interstellar Asteroid." *Nature* 552 (2017).

6   What do you think it would have been named? Something fun to think about as you cruise through it: J.-M. Petit, A. Morbidelli, and J. Chambers, "The Primordial Excitation and Clearing of the Asteroid Belt," (PDF), *Icarus* 153 (2001).

7   And of all the cans of worms you definitely never want to open, this just might be the biggest. Seriously, just don't even go there: Jean-Luc Margot, "A Quantitative Criterion for Defining Planets," *Astronomical Journal* 150 (2015).

8   It's always hard to pick a single paper to represent an entire field of research, because scientists are very jealous, but here's a good introduction to the wonders and mysteries of those adorable little frozen dirt balls (I'm talking about comets, not scientists): A. Morbidelli, "Origin and Dynamical Evolution of Comets and Their Reservoirs," Lecture on Comet Dynamics and the Outer Solar System, 35th Saas-Fee Advanced Course (2005),

9   Oh, and the "Oort" part is because of this: Jan Oort, "The Structure of the Cloud of Comets Surrounding the Solar System and a Hypothesis Concerning Its Origin," *Bulletin of the Astronomical Institutes of the Netherlands* 11 (1950).

10   And if you ever find yourself in too good a mood, this will do the trick: Bostrom, Nick, "Existential Risks: Analyzing Human Extinction Scenarios and Related Hazards", *Journal of Evolution and Technology*, 9 (2002)

SOLAR FLARES AND CORONAL MASS EJECTIONS

1   The actual chain reaction is much more complicated than that, involving a dance of four protons. Plus sometimes other elements can get in on the action. Still, nuclear physicists are smart folks and have it largely figured out: Christian Iliadis, *Nuclear Physics of Stars*, (Weinheim, Germany: Wiley-VCH, 2007).

2   I mean you really have to hand it to them. Astronomers can peer inside the heart of the sun without even attempting a vivisection (a solarsection?): C. J. Hansen, S. D. Kawaler, and V. Trimble, *Stellar Interiors*, (Berlin: Springer, 2004).

3   Just . . . sigh: Max Planck, "Über eine Verbesserung der Wien'schen Spectralgleichung," *Verhandlungen der Deutschen Physikalischen Gesellschaft* 2 (1900).

4   Of course everyone blames magnetic fields, because they're just sitting there looking all guilty and everything: Eric Priest, *Solar Magneto-hydrodynamics*, (Dorrecht, Holland: D. Reidel Publishing Company, 1982).

5   Maybe, sorta, kinda. It's the best thing we've got, so we're just going to stick with it for now, OK?: H. W. Babcock, 'The Topology of the Sun's Magnetic Field and the 22-Year Cycle," *Astrophysical Journal* 133 (1961).

6   You know when you played with toys as a kid and smashed things together randomly just to see what happened? This is the theoretical equivalent of that: Anthony R. Bell, "The Acceleration of Cosmic Rays in Shock Fronts—I," *Monthly Notices of the Royal Astronomical Society* 182 (1978).

7   Interestingly, Carrington (the guy we named the event after since he did such a wonderful job describing it) also noticed a massive flare on the sun in conjunction with the event. Coincidence? I think not: R. C. Carrington, "Description of a Singular Appearance seen in the Sun on September 1, 1859," *Monthly Notices of the Royal Astronomical Society* 20 (1859).

8   Want to know what the weather in space will be like this weekend? NASA's got you covered: "Space Weather," (https://www.nasa.gov/mission_pages/rbsp/science /rbsp-spaceweather.html), retrieved November 6, 2019.

9   Seriously this place is a mess: Brian E. Wood, Jeffrey L. Linsky, Hans-Reinhard Müller, and Gary P. Zank, "Observational Estimates for the Mass-Loss Rates of α Centauri and Proxima Centauri Using Hubble Space Telescope Lyα Spectra," *Astrophysical Journal* 547 (2001).

COSMIC RAYS

1   Well, not really. I suppose I should've said, you'll have to live with it. Or die with it: V. F. Hess, "Über Beobachtungen der durchdringenden Strahlung bei sieben Freiballonfahrten," *Physikalische Zeitschrift* 13 (1912).

2   I'm not in charge of names, I just work here: Anthony R. Bell, "The Acceleration of Cosmic Rays in Shock Fronts—I," *Monthly Notices of the Royal Astronomical Society* 182 (1978).

3   Ah, Fritz Zwicky. I don't get to talk about him nearly as much as I wish I could in these pages. Oh well, a footnote will have to do: W. Baade and F. Zwicky,

"Cosmic Rays from Super-novae," *Proceedings of the National Academy of Sciences of the United States of America* 20 (1934).

4   One of the first clues that something fishy was going on was the orientation of comet tails: they always point away from the sun: Lugwig Biermann, "Kometenschweife und solare Korpuskularstrahlung," *Zeitschrift für Astrophysik* 29 (1951).

5   As first discovered by the intrepid Voyager probes: J. S. Rankin, et al., "Heliosheath Properties Measured from a Voyager 2 to Voyager 1 Transient," *Astrophysical Journal* 883 (2019).

6   Now's a good time to pay respects to the Juno spacecraft, which sacrificed itself to give us so much knowledge: T. Gastine, et al., "Explaining Jupiter's Magnetic Field and Equatorial Jet Dynamics," *Geophysical Research Letters* 41 (2014).

7   And once you leave the galaxy, kiss all that wonderful magnetic protection goodbye: Philip Hopkins, et al., "But What About . . . Cosmic Rays, Magnetic Fields, Conduction, and Viscosity in Galaxy Formation," *Monthly Notices of the Royal Astronomical Society* (2019).

8   Oh, you think I'm joking, do you? Take this: Adrian L. Melott and Brian C. Thomas, "From Cosmic Explosions to Terrestrial Fires?" *Journal of Geology* 127 (2019).

9   UNSCEAR, "Sources and Effects of Ionizing Radiation," (http://www.unscear. org/docs/reports/2008/09-86753_Report_2008_Annex_B.pdf), retrieved November 6, 2019.

10 Copious literature on the subject abounds for the lonely traveler's amusement: M. Kachelriess and D. V. Semikoz, "Cosmic Ray Models," (arxiv.org/1904.08160), retrieved November 6, 2019.

11 I can totally see the comic-book-style "KA-BLAMO" that accompanied the event: D. J. Bird, et al., "Detection of a Cosmic Ray with Measured Energy Well Beyond the Expected Spectral Cutoff Due to Cosmic Microwave Radiation," *Astrophysical Journal* 441 (1995).

12 Nasty little buggers, too: C. Arguelles, et al., "Searches for Atmospheric Long-lived Particles," *Journal of High-Energy Physics* (2019).

13 This is some serious mad scientist-level stuff: G. de Wasseige, "High-energy Neutrino Astronomy: Current Status and Prospects," EPS-HEP 2019 Conference (2019).

14 Someday soon we'll rename computer technicians as electro-oncologists: T. Huckle and T. Neckel, *Bits and Bugs: A Scientific and Historical Review of Software Failures in Computational Science*, (Society for Industrial & Applied Mathematics, 2019).

STELLAR NURSERIES

1   We even have cute baby pictures of it: Planck Collaboration, "Planck 2018 Results. I. Overview and the Cosmological Legacy of Planck," *Astronomy & Astrophysics* (2018).

2   There's also a fair bit of dust, which we'll talk about more soon, ranging in size from "really small" to "also really small": J. S. Mathis, W. Rumpl, K. H. Nordsieck, "The Size Distribution of Interstellar Grains, *Astrophysical Journal* 217 (1977).

3   Not bok as in "chicken" but bok as in: Bart J. Bok and Edith F. Reilly, "Small Dark Nebulae," *Astrophysical Journal* 105 (1947).

4   Exactly how much of a cloud is really, truly available for star formation is a subject of intense, grueling study by astronomers and explorers alike: J. P. Williams, L. Blitz, and C. F. McKee, "The Structure and Evolution of Molecular Clouds: From Clumps to Cores to the IMF," *Protostars and Planets IV*, (Tucson: University of Arizona Press, 2000).

5   Like that great singer at the local bar that's just waiting to make it big: C. Hayashi, "The Evolution of Protostars," *Annual Review of Astronomy and Astrophysics* 4 (1966).

6   Give it up for the double-H: B. Reipurth and C. Bertout, eds., "50 Years of Herbig–Haro Research. From Discovery to HST," *Herbig–Haro Flows and the Birth of Stars; IAU Symposium No. 182*, (Dordrecht: Kluwer Academic Publishers, 1997).

7   As soon as an astronomer spots something new, they have two choices: name it after the location in the sky where they found it, or try to name it after themselves: Alfred H. Joy, "T Tauri Variable Stars," *Astrophysical Journal* 102 (1945).

8   I mean, somebody's gotta get the good genes in the family: N. Murray, "Star Formation Efficiencies and Lifetimes of Giant Molecular Clouds in the Milky Way." *Astrophysical Journal* 729 (2011).

9   Honestly, good riddance. We never liked them anyway: Vasilii V. Gvaramadze and Alessia Gualandris, "Very Massive Runaway Stars from Three-body Encounters," *Monthly Notices of the Royal Astronomical Society* 410 (2010).

10 Of course there are alternative models out there, but this is the "generally accepted until somebody has a much, much better idea" model: C. C. Lin and F. H. Shu, "On the Spiral Structure of Disk Galaxies," *Astrophysical Journal* 140 (1964).

11 I categorically reject all names for the new galaxy that come from a combination of Milky Way and Andromeda: Sangmo Tony Sohn, Jay Anderson, and Roeland van der Marel, "The M31 Velocity Vector. I. Hubble Space Telescope Proper-motion Measurements," *Astrophysical Journal* 753 (2012).

12 Others call it "an area of active research": J. E. Pringle, "Accretion Discs in Astrophysics," *Annual Review of Astronomy and Astrophysics* 19 (1981).

13 You know something hurts when you can't even stand up straight 5 billion years later: Jay T. Bergstralh, Ellis Miner, and Mildred Matthews, *Uranus*, (Tucson: University of Arizona Press, 1991).

14 It's called the "Nice Model," not because it's very nice, but because astronomers had the brilliant idea of hosting a conference in Nice, France, while debating this: A. Crida, "Solar System Formation." *Reviews in Modern Astronomy* 21 (2009).

STELLAR-MASS BLACK HOLES

1   I'm surprised that he survived staring into the abyss: Karl Schwarzschild, "Über das Gravitationsfeld eines Massenpunktes nach der Einsteinschen Theorie," Sitzungsber. Preuss. Akad. D. Wiss (1916).

2   You can thank Einstein's former teacher, Hermann Minkowski, for the realization that time and space are forever united: Hermann Minkowski, "Raum und Zeit, *Jahresbericht der Deutschen Mathematiker-Vereinigung* (1909).

3   It's true that nobody could do it quite like old Uncle Albert himself: Albert Einstein, "Die Feldgleichungen der Gravitation," *Sitzungsberichte der Preussischen Akademie der Wissenschaften zu Berlin* (1915).

4   It's even gotten to the point that some are arguing that maybe it's not even a problem at all. That's when you know you're really in the weeds: S. Fichet, "Quantified Naturalness from Bayesian Statistics," *Physical Review D.* 86 (2012).

5   Basically as soon as we turned on X-ray detectors, there were the black holes, screaming at us from the depths all along: S. Bowyer, et al., "Cosmic X-ray Sources," *Science* 147 (1965).

6   It's called "microlensing" because "gravitational tweaking" just sounded too absurd for Serious Scientific Discourse: Joachim Wambsganss, "Gravitational Microlensing," *Gravitational Lensing: Strong, Weak and Micro*, Saas-Fee Lectures, Springer-Verlag, Saas-Fee Advanced Courses 33 (2006).

7   The seventies were a trippy time, man: Stephen Hawking, "Black hole explosions?" *Nature* 248 (1974).

8   While for generations people had guessed that the tides had something to do with the sun and/or moon, it was old bigwig himself who solved it mathematically with his newfangled universal gravity: Isaac Newton, *Philosophiae Naturalis Principia Mathematica* (1729).

9   It was really coined by (a very hungry) Stephen Hawking in his best-selling book, and later propagated back into the (very hungry) science community: Stephen Hawking, *A Brief History of Time*, (New York: Bantam Dell, 1988).

10 You've got to hand it to the astronomers, though, given how hard they're working on hunting for them: A. Celotti, J. C. Miller, and D. W. Sciama, "Astrophysical Evidence for the Existence of Black Holes," *Classical and Quantum Gravity* 16 (1999).

PLANETARY NEBULAE

1   Is it getting hot in here, or is it just me?: Donald D. Clayton, *Principles of Stellar Evolution and Nucleosynthesis*, (Chicago: University of Chicago Press, 1983).

2   No, it's definitely not me: K.-P. Schröder and Robert Connon Smith, "Distant Future of the Sun and Earth Revisited," *Monthly Notices of the Royal Astronomical Society* 386 (2008).

3   Blessed are the red, for they will inherit the universe: Michael Richmond, "Late Stages of Evolution for Low-mass Stars," (http://spiff.rit.edu/classes/phys230 /lectures/planneb/planneb.html), retrieved November 8, 2019.

4   Honestly, who doesn't want to end their lives as a bloated red monstrosity?: I. -J. Sackmann, A. I. Boothroyd, and K. E. Kraemer, "Our Sun. III. Present and Future," *Astrophysical Journal* 418 (1993).

5 You might want to have a doctor check that out: Martin Schwarzschild, "On the Scale of Photospheric Convection in Red Giants and Supergiants," *Astrophysical Journal* 195 (1975).

6 You brought a Geiger counter, right?: R. G. Deupree and R. K. Wallace, "The Core Helium Flash and Surface Abundance Anomalies," *Astrophysical Journal* 317 (1987).

7 This whole chapter (and indeed the entire field of study) can be summarized as "and then it gets even worse": David R. Alves and Ata Sarajedini, "The Age-dependent Luminosities of the Red Giant Branch Bump, Asymptotic Giant Branch Bump, and Horizontal Branch Red Clump," *Astrophysical Journal* 511 (1999).

8 I think that star is winking at us: I. S. Glass and T. Lloyd Evans, "A Period-luminosity Relation for Mira Variables in the Large Magellanic Cloud," *Nature* 291 (1981).

9 I can just imagine it sounding like a very slow, deep, long *whoooooooooosh*: P. R. Wood, E. A. Olivier, and S. D. Kawaler, "Long Secondary Periods in Pulsating Asymptotic Giant Branch Stars: An Investigation of Their Origin," *Astrophysical Journal* 604 (2004).

10 A handy guide for the prudent explorer: David J. Frew, *Planetary Nebulae in the Solar Neighbourhood: Statistics, Distance Scale and Luminosity Function*, PhD thesis, Department of Physics, Macquarie University, Sydney, Australia (2008).

WHITE DWARVES AND NOVAE

1 Congratulations, you're now a quantum mechanic: David J. Griffiths, *Introduction to Quantum Mechanics* (2nd edition), (London, UK: Prentice Hall, 2005).

2 Paul Dirac will not appear in this book, sadly, except as a footnote to say that he suggested the names: Graham Farmelo, *The Strangest Man: The Hidden Life of Paul Dirac, Mystic of the Atom*, (New York: Basic Books, 2009).

3 The real guts of this is based on a deep and potentially insane connection between the spin of a particle and the kind of statistics that it obeys. Don't ask me: Wolfgang Pauli, "The Connection between Spin and Statistics," *Physical Review* 58 (1940).

4 Hup-hup-hooray for Heisenberg: W. Heisenberg, "Über den anschaulichen Inhalt der quantentheoretischen Kinematik und Mechanik," *Zeitschrift für Physik* 43 (1927).

5 Little stars, big thoughts: S. Chandrasekhar, "The Highly Collapsed Configurations of a Stellar Mass (Second paper)," *Monthly Notices of the Royal Astronomical Society* 95 (1935).

6 Yes, it's more complicated than that, because it's always more complicated than that: J. B. Holberg, "How Degenerate Stars Came to Be Known as White Dwarfs," *American Astronomical Society Meeting 207* 207 (2005).

7 To be fair, it seems that Sir Arthur Eddington was fond of telling many people and physical objects to shut up: A. S. Eddington, *Stars and Atoms* (Oxford, UK: Clarendon Press, 1927).

8   Don't touch it, it might break!: S. Chandrasekhar, "Stellar Configurations with Degenerate Cores," *Observatory* 57 (1934).

9   Just chilling out, astronomically speaking: Daniel J. Eisenstein, et al., "A Catalog of Spectroscopically Confirmed White Dwarfs from the Sloan Digital Sky Survey Data Release 4," *Astrophysical Journal Supplement Series* 167 (2006).

10   If you're willing to dig you might just expose one today: J. L. Barrat, J. P. Hansen, and R. Mochkovitch, "Crystallization of Carbon-oxygen Mixtures in White Dwarfs," *Astronomy & Astrophysics* 199 (1988).

11   Like fireworks. When they go off in the sky, you can ooh and ahh to your heart's content. When they go off at your feet . . . ouch: M. J. Darnley, et al., "On the Progenitors of Galactic Novae," *Astrophysical Journal* 746 (2012).

12   With a real gem of a book title: T. Brahe, *De nova et nullius ævi memoria prius visa stella*, (1572).

13   Now kiss: R. Tylenda, et al., "V1309 Scorpii: Merger of a Contact Binary," *Astronomy & Astrophysics* 528 (2011).

14   R. M. Crocker, et al., "Diffuse Galactic Antimatter from Faint Thermonuclear Supernovae in Old Stellar Populations," *Nature Astronomy* 1 (2017).

SUPERNOVAE

1   Thanks, unknown Chaco astronomer, you're a real pal: "Supernova Pictograph," (https://www2.hao.ucar.edu/Education/SolarAstronomy/supernova-pictograph), retrieved November 8, 2019.

2   And you too, Chinese astronomers: Zhentao Xu and David W. Pankenier, "East-Asian Archaeoastronomy: Historical Records of Astronomical Observations of China, Japan, and Korea," (2000).

3   I can just see Zwicky making this statement while artfully adjusting his bolo tie: W. Baade and F. Zwicky, "On Super-novae," *Proceedings of the National Academy of Sciences* 20 (1934).

4   You may select the verb of your choice: sizzling, popping, or bubbling: E. Cappellaro and M. Turatto, "Supernova Types and Rates," *Influence of Binaries on Stellar Population Studies* 264 (Dordrecht, Netherlands: Kluwer Academic Publishers, 2001).

5   Is there something wrong with the hearts of scientists who study how stars die? How do they sleep at night?: S. E. Woosley and H. -T. Janka, "The Physics of Core-Collapse Supernovae," *Nature Physics* 1 (2005).

6   And not just "some neutrinos" but "more neutrinos than strictly necessary": E. S. Myra and A. Burrows, "Neutrinos from Type II Supernovae: The First 100 Milliseconds," *Astrophysical Journal* 364 (1990).

7   Imagine shaking a soda bottle the size of a giant star: Wakana Iwakami, et al., "3D Simulations of Standing Accretion Shock Instability in Core-Collapse Supernovae," 14th Workshop on Nuclear Astrophysics (2008).

8   Boil, boil, toil and trouble: Paul A. Crowther, "Physical Properties of Wolf–Rayet Stars," *Annual Review of Astronomy and Astrophysics* 45 (2007).

9   Look in the mirror, do you see the shining star inside you?: Jennifer A. Johnson, "Populating the Periodic Table: Nucleosynthesis of the Elements," *Science* 363 (2019).

10 And don't think a pair of sunglasses will protect you: W. Hillebrandt and J. C. Niemeyer, "Type IA Supernova Explosion Models," *Annual Review of Astronomy and Astrophysics* 38 (2000).

11 We're just tossing out ideas, hoping one of them will stick: K. Nomoto, M. Tanaka, N. Tominaga, and K. Maeda, "Hypernovae, Gamma-ray Bursts, and First Stars," *New Astronomy Reviews* 54 (2010).

12 Imagine taking the pit out of an avocado and then the avocado exploding: Gary S. Fraley, "Supernovae Explosions Induced by Pair-Production Instability," (PDF), *Astrophysics and Space Science* 2 (1968).

NEUTRON STARS AND MAGNETARS

1   OK, I'm pretty sure Fritz Zwicky was trying to become some sort of cosmic super villain: Walter Baade and Fritz Zwicky, "Remarks on Super-Novae and Cosmic Rays," *Physical Review* 46 (1934).

2   And the going quickly gets weird: Paweł Haensel, Alexander Y. Potekhin, and Dmitry G. Yakovlev, *Neutron Stars*, (Berlin: Springer, 2007).

3   And you thought your shoulders were tight: B. Haskell and A. Melatos, "Models of Pulsar Glitches," *International Journal of Modern Physics D* 24 (2015).

4   Please remove all metal objects before approaching a magnetar: Victoria M. Kaspi and Andrei M. Beloborodov, "Magnetars," *Annual Review of Astronomy and Astrophysics* 55 (2017).

5   The pulses were so perfectly regular that the first object was dubbed "LGM-1" for "Little Green Men." You know, just in case the astronomers had been the first to discover alien life: A. Hewish, et al., "Observation of a Rapidly Pulsating Radio Source," *Nature* 217 (1968).

6   The combined observation of a kilonovae with both gravitational waves and the more familiar electromagnetic waves also had the added bonus of decapitating almost every theory of modified gravity out there: B. P. Abbott, et al., "Multi-messenger Observations of a Binary Neutron Star Merger," *Astrophysical Journal Letter* 848 (2017).

7   Well, how about it?: D. D. Ivanenko and D. F. Kurdgelaidze, "Hypothesis Concerning Quark Stars," *Astrophysics* 1 (1965).

SUPERMASSIVE BLACK HOLES

1   There are also some really great jazz clubs: W. Baade, "A Search for the Nucleus of Our Galaxy," *Publications of the Astronomical Society of the Pacific* 58 (1946).

2   Looking back, what a missed opportunity to give it a truly awe-inspiring name. Oh well, maybe next time: B. Balick and R. L. Brown, "Intense Sub-arcsecond Structure in the Galactic Center," *Astrophysical Journal* 194 (1974).

3   So much so that there appears to be a relationship between the size of the central black hole and the properties of the galaxy itself. It seems as if the galaxies can't grow without their dark secrets: K. Gebhardt, et al., "A Relationship Between Nuclear Black Hole Mass and Galaxy Velocity Dispersion," *Astrophysical Journal* 539 (2000).

4   Told you so: Miloš Milosavljević and David Merritt, "The Final Parsec Problem," (PDF), AIP Conference Proceedings, American Institute of Physics, 686 (2003).

5   To be more precise (because why not), there are a few candidate medium-sized black holes, but naturally they're hard to pin down observationally: Karl Gebhardt, R. M. Rich, and Luis C. Ho, "An Intermediate-Mass Black Hole in the Globular Cluster G1: Improved Significance from New Keck and Hubble Space Telescope Observations," *Astrophysical Journal* 634 (2005).

6   Abstract, simplified: "Looks like we've got nothin'": A. Ghez, et al., "High Proper-Motion Stars in the Vicinity of Sagittarius A*: Evidence for a Supermassive Black Hole at the Center of Our Galaxy," *Astrophysical Journal* 509 (1998).

7   Everybody say "Schwarzschild!": The Event Horizon Telescope Collaboration, "First M87 Event Horizon Telescope Results. I. The Shadow of the Supermassive Black Hole," *Astrophysical Journal Letters* 87 (2019).

8   "Experienced" as in "the quantum nature of reality rips you to smithereens": Ahmed Almheiri, et al., "Black Holes: Complementarity or Firewalls?" *Journal of High Energy Physics* 2013 (2013).

9   Just don't look the laser in the eye: R. V. Pound and G. A. Rebka Jr., "Gravitational Red-Shift in Nuclear Resonance," *Physical Review Letters* 3 (1959).

10 Take my hand and go on a journey with me, a journey you won't regret (because you won't live to even remember it): "Journey into a Schwarzschild Black Hole," (https://jila.colorado.edu/~ajsh/insidebh/schw.html), retrieved November 8, 2019.

11 If something is about to fall into a black hole, just let it go, man: J. A. Wheeler, Pontificae Acad. Sei. Scripta Varia, 35, 539 (1971).

QUASARS AND BLAZARS

1   Oh look at all those pretty blinking lights! Sigh, we were so naive: Gregory A. Hields, "A Brief History of AGN," *Publications of the Astronomical Society of the Pacific* 111 (1999).

2   And conveniently enough they're called the Brightest Cluster Galaxies: Y-T. Lin and Joseph J. Mohr, "K-band Properties of Galaxy Clusters and Groups: Brightest Cluster Galaxies and Intracluster Light," *The Astrophysical Journal* 617 (2004).

3   Why does nature insist on outdoing herself, time and again?: Tiziana Di Matteo, et al., "Energy Input from Quasars Regulates the Growth and Activity of Black Holes and Their Host Galaxies," *Nature* 433 (2005).

4   Just wind it up and let it rip: Eric G. Blackman, "Accretion Disks and Dynamos: Toward a Unified Mean Field Theory," Proceedings of the Third International Conference and Advanced School, "Turbulent Mixing and Beyond," (2012).

5   Perhaps most amazingly, we figured this out while black holes were purely
    hypothetical objects: H. Thirring, "Über die Wirkung rotierender ferner Massen
    in der Einsteinschen Gravitationstheorie," *Physikalische Zeitschrift* 19 (1918).

6   For example, English: R. Penrose and R. M. Floyd, "Extraction of Rotational
    Energy from a Black Hole," *Nature Physical Science* 229, 177 (1971).

7   Giant. Space. Lasers. Need I say more?: R. Blandford, et al., "Relativistic Jets from
    Active Galactic Nuclei," *Annual Review of Astronomy and Astrophysics* 57 (2019).

8   Giving us some of the most frighteningly beautiful structures in the entire
    universe: R. C. Jennison and M. K. Das Gupta, "Fine Structure of the Extra-
    Terrestrial Radio Source Cygnus 1," *Nature* 172 (1953).

9   I think nature is just messing with us now: M. Ruszkowski, et al., "Impact of
    Tangled Magnetic Fields on Fossil Radio Bubbles," *Monthly Notices of the Royal
    Astronomical Society* 378 (2007).

10 You drive a hard bargain, universe: Tiziana Di Matteo, et al., "Energy Input
    from Quasars Regulates the Growth and Activity of Black Holes and Their Host
    Galaxies," *Nature* 433 (2005).

11 Don't ever wish for a time machine, the young universe was even worse than the
    one we know today: Ryle and Clarke, "An Examination of the Steady-state Model
    in the Light of Some Recent Observations of Radio Sources," *MNRAW* 122
    (1961).

COSMIC STRINGS AND MISCELLANEOUS SPACETIME DEFECTS

1   An old, ancient universe, full of terrors through the ages: The Planck
    Collaboration, "Planck 2018 Results. I. Overview, and the Cosmological Legacy of
    Planck," *Astronomy & Astrophysics* (2018).

2   If you close your eyes you can almost feel all the fields wiggling through you:
    M. Peskin and D. Schroeder, *An Introduction to Quantum Field Theory*, (Boulder,
    Colo.: Westview Press, 1995).

3   They need a catchphrase when squaring off with the monster of the week:
    H. Georgi and S. L. Glashow, "Unity of All Elementary Particle Forces," *Physical
    Review Letters* 32 (1974).

4   If you've ever wondered why the search for the Higgs boson was such a big deal, it's
    because the Higgs is the guilty party for creating this split: Peter W. Higgs, "Broken
    Symmetries and the Masses of Gauge Bosons," *Physical Review Letters* 13 (1964).

5   Come on folks, hurry up already so we can go home: G. Ross, *Grand Unified
    Theories*, (Boulder, Colo.: Westview Press, 1984).

6   And you thought your dorm rooms were cramped: Jonathan Allday, *Quarks,
    Leptons and the Big Bang,* (Bristol, England: Institute of Physics Publishing, 2001).

7   I mean, just look at how banal this title is: G. Hooft, "Magnetic Monopoles in
    Unified Gauge Theories," *Nuclear Physics B* 79 (1974).

8   At least, not any monopole that has made itself glaringly obvious: Blas
    Cabrera, "First Results from a Superconductive Detector for Moving Magnetic
    Monopoles," *Physical Review Letters* 48 (1982).

9   Quit making a mountain out of a molehill, universe: Alan Guth, "Inflationary Universe: A Possible Solution to the Horizon and Flatness Problems," *Physical Review D* 23 (1981).

10 And once broken, the universe can never again heal: Edmund J. Copeland, et al., "Cosmic F- and D-strings," *Journal of High Energy Physics* 2004 (2004).

11 Why tease us, universe, why?: Zaven Arzoumanian, et al., "The NANOGrav Nine-year Data Set: Limits on the Isotropic Stochastic Gravitational Wave Background," *Astrophysical Journal* 821 (2015).

12 It's a real head-scratcher. By all rights our universe probably shouldn't even exist. And yet . . . here we are. What gives?: N. Straumann, "Cosmological Phase Transitions," Invited lecture at the third Summer School on Condensed Matter Research, (2004).

DARK MATTER

1   And here comes Fritz Zwicky again in the footnotes, as he was the one to name it: F. Zwicky, "Die Rotverschiebung von extragalaktischen Nebeln," *Helvetica Physica Acta* 6 (1933).

2   Fine. Four percent of all the matter and energy in the universe: L. Bergstrom, "Non-baryonic Dark Matter: Observational Evidence and Detection Methods," *Reports on Progress in Physics* 63 (2000).

3   If anyone was robbed of a Nobel Prize, it was probably Vera Rubin: V. Rubin, W. K. Thonnard Jr., and N. Ford, "Rotational Properties of 21 Sc Galaxies with a Large Range of Luminosities and Radii from NGC 4605 (R = 4kpc) to UGC 2885 (R = 122kpc)," *Astrophysical Journal* 238 (1980).

4   F. Zwicky, "Die Rotverschiebung von extragalaktischen Nebeln," *Helvetica Physica Acta* 6 (1933).

5   Folks, I think the universe is trying to tell us something: A. Refregier, "Weak Gravitational Lensing by Large-scale Structure," *Annual Review of Astronomy and Astrophysics* 41 (2003).

6   We can always change Mercury's name, regardless of what Albert says: Albert Einstein, "The Foundation of the General Theory of Relativity," *Annalen der Physik* 49 (1916).

7   Seriously, how many humans can claim to have found a new planet in the solar system?: N. Kollerstrom, "A Neptune Discovery Chronology," The British Case for Co-prediction, University College London, (2001).

8   Ultimately wrong but it was worth a shot: M. Milgrom, "A Modification of the Newtonian Dynamics as a Possible Alternative to the Hidden Mass Hypothesis," *Astrophysical Journal* 270 (1983).

9   Here's a hint: it rhymes with "mark hatter": Planck Collaboration, "Planck 2018 results. I. Overview and the Cosmological Legacy of Planck," *Astronomy & Astrophysics* (2018).

10 Whatever the universe is cooking, it smells great: R. A. Alpher, H. Bethe, and G. Gamow, "The Origin of Chemical Elements," *Physical Review* 73 (1948).

11   Welcome to theoretical physics, where all your wildest dreams can come true: Katherine Garrett and Gintaras Dūda, "Dark Matter: A Primer," *Advances in Astronomy* (2011).

12   It was either neutrinos exist or conservation of momentum is broken. We went with Door #1: Laurie M. Brown, "The Idea of the Neutrino," *Physics Today* 31 (1978).

13   And cupcakes wouldn't exist either!: Matteo Viel, et al., "Constraining Warm Dark Matter Candidates including Sterile Neutrinos and Light Gravitinos with WMAP and the Lyman-αforest," *Physical Review D. American Physical Society (APS)* 71 (2005).

14   Or the violently named Bullet Cluster: Douglas Clowe, et al., "A Direct Empirical Proof of the Existence of Dark Matter," *Astrophysical Journal Letters* 648 (2006).

15   The universe might be weirder than we thought (imagine that): G. Bertone, "The Moment of Truth for WIMP Dark Matter," *Nature* 468 (2010).

16   At this point the book would be shorter if I listed the things that don't emit gamma rays: G. Bertone and D. Merritt, "Dark Matter Dynamics and Indirect Detection," *Modern Physics Letters A* 20 (2005).

17   Solve this riddle and you might just unlock the deepest mysteries of the universe— or just open up new ones: W. J. G. de Blok, "The Core-cusp Problem," *Advances in Astronomy* (2010).

HOSTILE ALIENS

1   The ultimate size of the cosmos comes down to this really funky period of history called inflation. There may even be more than one headache . . . I mean, universe: Alan H. Guth, "Inflationary Universe: A Possible Solution to the Horizon and Flatness Problems," *Physical Review D* 23 (1981).

2   Using schoolchildren to prove their points: J. E. Evans and E. W. Maunder, "Experiments as to the Actuality of the 'Canals' Observed on Mars," *Monthly Notices of the Royal Astronomical Society* 63 (1903).

3   And if you've been reading the footnotes like the good little scholar that you are, you already knew this: A. Hewish, et al., "Observation of a Rapidly Pulsating Radio Source," *Nature* 217 (1968).

4   Various nonalien explanations have been put forth over the decades, some more convincing than others. But there is one thing that everyone has agreed on: something that interesting could have never happened in Columbus, Ohio: Robert Gray and S. Ellingsen, "A Search for Periodic Emissions at the Wow Locale," *Astrophysical Journal* 578 (2002).

5   Another reason why a truly civilized person waits for the microwave to ding: Petroff, E., et al., "Identifying the Source of Perytons at the Parkes Radio Telescope," *Monthly Notices of the Royal Astronomical Society* 451 (2015).

6   Astronomers around the world knew it was dust all along, because dust is responsible for almost every single anomalous signal in astronomy: Tabetha S. Boyajian, et al.,"The First Post-Kepler Brightness Dips of KIC 8462852," *Astrophysical Journal* 853 (2018).

7   At least it wasn't dust: K. J. Meech, et al., "A Brief Visit from a Red and Extremely Elongated Interstellar Asteroid," *Nature* 552 (2017).

8   Find the nearest geologist, hug them, and tell them that rocks are cool too: Lars Borg, et al., "The Age of the Carbonates in Martian Meteorite ALH84001," *Science* 286 (1999).

9   And every year the debate continues: should we bother to keep listening? Maybe the aliens are there, but they're just not that into us: "SETI at 50," *Nature* 416 (2009).

10   One word: rollercoasters: Freeman J. Dyson, "Search for Artificial Stellar Sources of Infra-Red Radiation," *Science* 131 (1960).

11   When arguments are simply circling around each other for decades, you start to wonder if we're even asking the right questions: Michael H. Hart, "Explanation for the Absence of Extraterrestrials on Earth," *Quarterly Journal of the Royal Astronomical Society* 16 (1975).

12   In the end, it looks like the Drake equation is a machine that turns hopes and dreams into printable numbers: T. L. Wilson, "The Search for Extraterrestrial Intelligence," *Nature* 409 (2001).

13   Don't lick Martian puddle water, it's poisonous: L. Ojha, et al., "Spectral Evidence for Hydrated Salts in Recurring Slope Lineae on Mars," *Nature Geoscience* 8 (2015).

14   Sometimes it even squirts out into space through one of my favorite astronomy words out there: a cryovolcano: L. Roth, et al., "Transient Water Vapor at Europa's South Pole," *Science* 343 (2013).

15   Robert Brown, Jean Pierre Lebreton, and Jack Waite, eds. *Titan from Cassini-Huygens*, (Berlin: Springer Science and Business Media 2009).

16   Only if we also get a mug of hot cocoa: John Johnson, *How Do You Find an Exoplanet?* (Princeton, N.J.: Princeton University Press 2015).

17   And it's even in the Habitable Zone of that little red star, an intriguing state of affairs to say the least: G. Anglada-Escudé, et al., "A Terrestrial Planet Candidate in a Temperate Orbit around Proxima Centauri," *Nature* 536 (2016).

18   That number is a real ballpark, so don't hang your hat on that quite yet, as much as you'd like to stake a claim to one of those lovely little worlds: Eric A. Petigura, et al., "Prevalence of Earth-size Planets Orbiting Sun-like Stars," *Proceedings of the National Academy of Sciences of the United States of America* 110 (2013).

19   Take a gander yourself: https://breakthroughinitiatives.org/initiative/3.

20   Even though everyone immediately recognized the correctness of special relativity, nobody wanted it to be true, because it was so weird: Edwin F. Taylor and John Archibald Wheeler, *Spacetime Physics: Introduction to Special Relativity*, (New York: W. H. Freeman, 1992).

WORMHOLES

1   Like almost every other amazing consequence from relativity, it was worked out almost immediately after Einstein published his work: Ludwig Flamm, "Beiträge zur Einsteinschen Gravitationstheorie," *Physikalische Zeitschrift* XVII: 448 (1916).

2   And like almost every other amazing consequence from relativity, it took decades to work out the full implications; R. W. Fuller and J. A. Wheeler, "Causality and Multiply-Connected Space-Time," *Physical Review* 128 (1962).

3   The very German math: H. Reissner, "Über die Eigengravitation des elektrischen Feldes nach der Einsteinschen Theorie," *Annalen der Physik* 50 (1916).

4   Full disclosure: I would buy a ticket for that ride: Roy P. Kerr, "Gravitational Field of a Spinning Mass as an Example of Algebraically Special Metrics," *Physical Review Letters* 11 (1963).

5   Don't blame me, I'm not responsible for the current state of the universe: Michael Morris, Kip Thorne, and Ulvi Yurtsever, "Wormholes, Time Machines, and the Weak Energy Condition," *Physical Review Letters* 61 (1988).

6   It does get much trickier and more nuanced here (which is probably true for this entire book), but here's a good starting point: H. Bondi, "Negative Mass in General Relativity," *Reviews of Modern Physics* 29 (1957).

7   Oh, it's real alright: S. K. Lamoreaux, "Demonstration of the Casimir Force in the 0.6 to 6 μm Range," *Physical Review Letters* 78 (1997).

8   Who knew that this conversation could so . . . academic: Christopher J. Fewster, Ken D. Olum, and Michael J. Pfenning, "Averaged Null energy Condition in Spacetimes with Boundaries," *Physical Review D* 75 (2007).

9   They end up in the Land. Wow, That's a Lot of Math: Z. Fu, et al., "Traversable Asymptotically Flat Wormholes with Short Transit Times," (arXiv e-print: 1908.03273).

INDEX